On Determination of the Degree of Dissociation of

Hydrogen in Non-Equilibrium Plasmas

by Means of Emission Spectroscopy.

Inaugural dissertation

zur

Erlangung des akademischen Grades

doctor rerum naturalium (Dr. rer. nat.)

an der Mathematisch-Naturwissenschaftlichen Fakultät

der

Ernst-Moritz-Arndt-Universität Greifswald

Vorgelegt von

Andrei V. Pipa

geboren am 31. Januar 1977

in Murmansk

Greifswald, 22. Juni 2004

Bibliografische Information Der Deutschen Bibliothek

Die Deutsche Bibliothek verzeichnet diese Publikation in der Deutschen
Nationalbibliografie; detaillierte bibliografische Daten sind im Internet über
http://dnb.ddb.de abrufbar.

ISBN 3-8325-0645-4

Logos Verlag Berlin
Comeniushof, Gubener Str. 47,
10243 Berlin
Tel.: +49 030 42 85 10 90
Fax: +49 030 42 85 10 92
INTERNET: http://www.logos-verlag.de

Dekan: Prof. Dr. J.-P. Hildebrandt

1. Gutachter: PD Dr. J. Röpcke

2. Gutachter: Prof. Dr. B.P. Lavrov

Tag der Promotion: 11. Oktober 2004

Content

1.   INTRODUCTION

The degree of dissociation of molecular hydrogen is considered as one of the most important parameters of hydrogen and hydrogen-containing low-temperature plasmas, which are of widespread interest in basic research, industrial applications and also in understanding of outer space phenomena. These plasmas are usually far from an equilibrium. The degree of dissociation is defined as the ratio of the number densities of dissociated molecules and molecules in all states, as ground, excited, dissociated and ionized states. Thus for a hydrogen plasma with negligible concentrations of excited and ionized species the degree of dissociation can be expressed as

$$D = \frac{[H]}{[H] + 2[H_2]} = \frac{[H]/[H_2]}{[H]/[H_2] + 2},$$

where [H] is the number density of atomic hydrogen in the ground $1^2S_{1/2}$ state, and [H$_2$] is the that for the hydrogen molecules in the $X^1\Sigma_g^+$ electronic ground state, summarized over all vibrational and rotational levels.

In the case of a thermodynamic equilibrium the degree of dissociation is a known function of temperature and pressure, while at non-equilibrium conditions, which are typical for low-temperature plasmas, a calculation of such a characteristics is usually extremely difficult. The main reason for those difficulties is the fact, that all important elementary processes, as e.g. dissociation, diffusion, association, and certain boundary conditions should be included in a theoretical model.

Sometimes, in the cases of a simple geometry it is possible to develop a relatively realistic model, which can be useful for application. For example, in a monoplasmatron ion source the degree of dissociation of hydrogen could be calculated in the plasma near to the anode [1]. This calculation was based on a certain kinetic model using measured space distributions of the electron energy distribution function and of the gas temperature. The effective value of the coefficient of surface association was treated as an adjusted parameter. But in general, even a semi-empiric way of calculations leads to tremendous difficulties in the majority of plasma systems caused by the uncertainties of rate coefficients of volume and surface processes and by sophisticated boundary conditions.

During the last two decades, several experimental methods for the determination of the degree of dissociation have been proposed, realized and used for studies of hydrogen or hydrogen-containing low-temperature plasmas. All of them have their own advantages and disadvantages.

It should be noted that, chemical and calorimetric methods are in principle non-local and indirect [2]. Measurements of the Lyman-α absorption in the vacuum ultra violet (VUV) spectral range need a rather special spectroscopic technique, which causes additional experimental difficulties in the majority of the applications of the gas discharges in particular in plasma technology [3]. Attractive laser methods as laser induced fluorescence (LIF), resonance enhanced multi-photon ionization (REMPI) and coherent anti-Stokes Raman scattering (CARS) also need expensive and complicate equipment. In addition, these methods have serious problems with absolute particle density calibration and secondary effects [4-6].

Recently it was shown that absolute intensity values of the H_2 continuum can be used for spectroscopic measurements of the rate of radiative dissociation via the $b^3\Sigma_u^+$ repulsive state and for rough estimations of the degree of dissociation of hydrogen, in the case the life time of atoms is known [7].

In low-temperature hydrogen and hydrogen-containing plasmas it is relatively easy to measure intensities of several spectral lines of the Balmer series (H_α, H_β, H_γ...). These lines correspond to spontaneous emissions of hydrogen atoms due to transitions between upper states with principle quantum numbers of n= 3, 4, 5... and of n = 2 of the lower state. The intensities of these lines contain information about population densities of excited levels of atomic hydrogen. These lines could be used for the determination of ground state population densities of atoms or of the degree of dissociation applying a suitable excitation-deactivation model.

The most direct method of spectroscopic determination of the degree of dissociation of hydrogen from intensities of Balmer lines has been first reported in [8, 9]. These studies were based on a detailed analysis of the excitation-deactivation balance equations, measured absolute line intensities, gas temperature, pressure and non-Maxwellian electron energy distribution functions (EEDF) obtained as a second derivative of Langmuir probe current-voltage characteristics [9]. In this study it was shown that at least for interpretation of the H_α line intensity the electron impact excitation from the ground state of the molecules, the dissociative excitation, as well as from the ground state of the atoms, the direct excitation, should be taken into account. The trapping of resonance radiation was considered as a change of the effective radiative lifetimes. The main advantages of this approach compared to other methods developed later [10-13] are: (i) the possibility to consider various secondary excitation, as stepwise excitation, dissociative recombination, etc. and deactivation processes, collisional quenching, stepwise ionization, etc; and (ii) its ability to work with an arbitrary shape of the EEDF which is known to be far from Maxwellian in hydrogen-containing discharge plasmas [9]. The necessity of absolute spectroscopic and Langmuir probe measurements is its main disadvantage.

The measurement procedure is significantly simplified in methods, which are based on relative line intensities. A ratio of intensities does not depend on the size of the plasma, if the plasma is optically thin, and the method does not need any absolute calibration of a detection system. In the case spectral lines have similar excitation cross sections then their intensity ratio does not depend considerably on the EEDF. Small differences in excitation thresholds and in the energy dependence of cross sections lead to weak dependencies of the intensity ratio on the shape of the EEDF. A simple approximation of the EEDF with few parameters allows to apply a pure spectroscopic diagnostics, based on measurements of several intensity ratios without the measurement of the EEDF.

One of such methods is actinometry [14]. Actinometry is applied to determine the density of atomic hydrogen. It is based on the measurement of the intensity ratios of Balmer lines and lines of neutral atoms of noble gases. For this purpose a rare gas added to hydrogen containing plasmas in small and known quantities. On one hand this admixture should be small enough to prevent disturbances of main characteristics of the plasma under the study. On the other hand, it should be high enough to be measurable with a sufficient precision governed by the intensity of the radiation. In particular, the two latter demands are obviously in contradiction, and it is not trivial to fulfill both simultaneously.

The other problem of actinometry is related to difficulties in involving the dissociative excitation of Balmer lines, what reduces significantly the range of the applicability of the method. Moreover actinometry needs an accurate consideration of the kinetic processes for noble gases as well as for hydrogen, what requires additional assumptions in related excitation-deactivation models.

The method to determine the degree of dissociation of hydrogen from the ratio of the H_α line intensity and the total intensity of diagonal Fulcher-α bands is described in [10]. In this work the vibro-rotational structure of levels and lines of the hydrogen molecule was ignored without any discussion while total emission cross sections of the bands obtained in gas-beam experiments, which are not valid for plasma conditions, were used in [10]. This drawback was later passed over into [11] using the intensity of single rovibronic molecular lines, namely the Q1 line of the (2-2) band of the Fulcher-α band system, the $d^3\Pi_u^-$, v = 2, N = 1 $\rightarrow$ $a^3\Sigma_g^+$, v = 2, N = 1 electronic-vibro-rotational transition. The EEDF in [10] is assumed to be Maxwellian and probe measurements of the electron temperature are supposed.

The determination of the degree of dissociation from relative intensities of Balmer lines was proposed and realized in [12]. This method is based on the assumption that two channels of excitation, direct and dissociative, have different relative contributions to the excitation of levels with different principal quantum numbers n. This method can be used in plasmas with a low degree of dissociation when the contributions of the two channels involved are not too different. The rate coefficients were found assuming a Maxwellian EEDF, but the authors treated a temperature of the Maxwellian function as adjusted parameter, which describes a real EEDF in the high-energy region above the excitation threshold. This assumption is more suitable for low-temperature hydrogen containing plasmas than the assumption of a Maxwellian EEDF. Since only the high-energy tail of the EEDF above 12 eV is responsible for the excitation of the spectral line, while most of the electrons have a mean energy usually at about 1 – 2 eV.

The Balmer lines are known to have an extremely narrow multiplet structure, the so-called fine structure of H I lines and levels. Doppler, instrumental and Stark broadenings are usually comparable or higher than the fine structure splitting. Therefore, in most experiments the total intensities of Balmer lines with unresolved fine structure are measured. The influence of an unresolved fine structure of Balmer lines on determining the degree of dissociation have been never discussed in literature as far it is known.

Recently [15] it was shown that the consideration of the fine structure in balance equations is very important for the calculation of emission rate coefficients and emission cross sections of direct and dissociative excitation of H_α and H_β lines. Also it was noticed that now there is not enough information available in order to use other Balmer lines for plasma diagnostics.

The main objective of the present work is the development of a new spectroscopic method for the determination of the degree of dissociation of hydrogen. Based on former approaches [10-12] this new method considers for the first time the non-resolved fine structure of Balmer lines in the kinetic of excitation and deactivation of hydrogen atoms.

A first report of the method has been made in [13].

The present work is structured in two main chapters, a theoretical and an experimental one.

In the theoretical chapter:

(i)   excitation processes of selected atomic and molecular spectroscopic lines of hydrogen are analyzed and a collision – radiative model is formulated,

(ii)   data in literature about relevant constants of elementary processes are compiled and the most reliable set of excitation cross sections and rate coefficients are presented for the first time,

(iii)   and finally a numerical experiment is performed to study range of applicability

The experimental chapter includes:

(i)   a description of experimental set-ups, where technical details of the plasma sources and of the spectroscopic equipment are given,

(ii)   an analysis of measured ratios of relevant hydrogen lines compared to the theoretical predictions,

(iii)   Approvement of the new method in three types of non-equilibrium discharges containing hydrogen:

    a) a capillary DC-arc discharge,

    b) a RF discharge, $f = 200$ kHz ,

    c) a MW discharge, $f = 2.45$ GHz.

The results of the present work show an opportunity of pure spectroscopic detection of the degree of dissociation of hydrogen in low temperature plasmas and indicate an insufficient knowledge about elementary processes for plasma diagnostics.

2. THEORETICAL MODEL AND ANALYSIS OF ELEMENTARY PROCESSES

2.1. Description of the excitation deactivation model

2.1.1. Excitation of Balmer lines

Two processes dominate in populating excited levels of hydrogen atoms in a wide range of plasmas conditions [9, 16]

i) The direct electron impact excitation of atom from the ground state

$$H\left(1^2 S_{1/2}\right)+e \rightarrow H^*(n,l,j)+e \tag{2.1.1},$$

where initial and final states are shown in brackets, n, l, j are principle, orbital and total angular momentum quantum numbers, and

ii) the dissociative electron impact excitation

$$H_2\left(X^1\Sigma_g^+, v, N\right)+e \rightarrow H^*(n,l,j)+H\left(1S_{1/2}\right)+e \tag{2.1.2},$$

in processes, where $X^1\Sigma_g^+$ is the electronic ground state of H_2 and v and N are the vibrational and rotational quantum numbers. The small triplet splitting of the levels can be neglected, and therefore the quantum number N of the total angular momentum, without considering the electron spin, is used as the characteristic feature of the rotational levels. For the deactivation of atoms the spontaneous emission is supposed to be most important [9, 16].

Taking into account these circumstances, the balance equation for a separate n-, l-, j-sublevel of an excited atom can be written as

$$[H]n_e\,K_{dir}^{ex}\left(F(\varepsilon)|nlj \leftarrow 1s\right)+[H_2]n_e\,K_{dis}^{ex}\left(F(\varepsilon)|nlj \leftarrow X\right)+R_{nlj}^+ = \frac{N_{nlj}}{\tau_{nlj}}+R_{nlj}^-, \tag{2.1.3}$$

where n_e is the electron concentration; $K_{dir}^{ex}\left(F(\varepsilon)|n,l,j < 1s\right)$ and $K_{dis}^{ex}\left(F(\varepsilon)|n,l,j \leftarrow X\right)$ are the excitation rate coefficients for the reactions (2.1.1) and (2.1.2); [H] and [H_2] are the concentrations of hydrogen atoms and molecules; N_{nlj} and τ_{nlj} are the population and the radiative lifetime of the nlj-sublevel; R_{nlj}^+ and R_{nlj}^- are the total rates of population and depopulation of a considered sublevel of the fine structure due to all other processes that can be referred as secondary.

In a strict sense, the excitation cross sections and the corresponding rate coefficients (2.1.2) can depend on the vibrational (v) and rotational (N) quantum numbers of the molecule's initial state. Unfortunately, nothing is known about it. The dissociative excitation was only studied by exciting molecular hydrogen using an electron beam in gas-filled electron guns. The gas pressure and current density in such experiments are usually lowered as far as the sensitivity of the detecting system it allows. Hence, it can be assumed that most of the molecules are in the vibrational ground state $X^1\Sigma_g^+$, $v = 0$, and the rotational and gas temperatures are rather low, i.e. in a range of 400–600 K.

In plasma experiments, high rotational and vibrational levels are populated due to gas heating and electron-impact vibrational excitation. Therefore, the rotational and gas temperatures in plasma are usually higher, 600–2000 K [9, 17], than in electron guns, while the vibrational temperature is 2500–5000 K [7]. To employ the experimentally determined cross sections in an analysis of plasma processes, one should assume that

(i) the cross sections of dissociative excitation are independent on N and

(ii) the concentration of vibrationally excited molecules in the states $X^1\Sigma_g^+$, $v \geq 1$ in a plasma is much lower than that in the ground vibronic state $X^1\Sigma_g^+$, $v = 0$.

Then the rate of dissociative excitation reaction can be considered as proportional to the total concentration [H_2] of molecules.

The rate coefficient of a certain b ← a transition is related to the cross section $\sigma_{b\leftarrow a}^{ex}(\varepsilon)$ of the electron-impact excitation of the level and the electron energy distribution function (EEDF) in a usual way:

$$K^{ex}(F(\varepsilon)|\, b \leftarrow a) = \int_{\varepsilon_{ab}}^{\infty} \sigma_{b\leftarrow a}^{ex}(\varepsilon) \sqrt{\frac{2\varepsilon}{m}} F(\varepsilon) d\varepsilon, \qquad (2.1.4)$$

where a and b is the set of quantum numbers characterizing the initial and final states, ε and m are the electron energy and mass, ε_{ab} is the threshold energy for the $b \leftarrow a$ process, and $F(\varepsilon)$ is the EEDF normalized as

$$\int_0^\infty F(\varepsilon) d\varepsilon = 1.$$

In the case of a Maxwellian EEDF, the rate coefficients become a function of only one parameter, i.e. of the electron temperature T_e:

$$K^{ex}(T_e|\, b \leftarrow a) = \frac{6.69 \cdot 10^7}{T_e^{\frac{3}{2}}} \int_{\varepsilon_{ab}}^{\infty} \sigma_{b\leftarrow a}^{ex}(\varepsilon)\varepsilon \exp\left(-\frac{\varepsilon}{T_e}\right) d\varepsilon \qquad (2.1.5)$$

where T_e, the excitation cross section, and the rate coefficient are given in eV, cm^2, and cm^{3}s^{-1}, respectively.

The intensity $I_{2\tilde{l}\tilde{j}}^{nlj}$ of a separate component of the fine structure of the Balmer line is expressed via the n, l, j-sublevel population, and the spontaneous emission probability (Einstein coefficient) for a $nlj \rightarrow 2\tilde{l}\ \tilde{j}$ transition is given by

$$I_{2\tilde{l}\tilde{j}}^{nlj} = N_{nlj}\ A_{2\tilde{l}\tilde{j}}^{nlj},$$ (2.1.6)

where $\tilde{l} = l \pm 1$, $\tilde{j} = j$ and $j \pm 1$ are taken according to the selection rules for electric dipole transitions. The intensity as the number of quanta emitted due to the transition considered is given per unit volume and per unit time in all directions.

The probabilities of radiative transitions of the hydrogen atom, calculated within the Schrödinger and Dirac theories, have the same numerical values [18]. Numerical data are listed in the appendix, table 1. The numerical data show the probabilities of spontaneous transitions, summed over $\tilde{j}$ independent on j, while radiative lifetimes depend only on the principal and orbital quantum numbers,

$$\sum_{j} A_{2\tilde{l}\tilde{j}}^{nlj} = A_{2\tilde{l}}^{nl}, \qquad \tau_{nlj} = \tau_{nl}.$$ (2.1.7)

In some cases, where j may be excluded from balance equations (2.1.3), this property of the hydrogen atom allows to consider sublevel pairs with identical quantum numbers n and l as a single nl-sublevel of the fine structure, which significantly simplifies the formulas.

The relations (2.1.7) can be indirectly confirmed by the fact that the measured and calculated lifetimes τ_{nl} are in a good agreement [19 – 22]. The consideration of nl-sublevels (1s, 2p, 3d, etc.) of the hydrogen atom is widely used in studies of radiative lifetimes, as well as in calculations and measurements of the excitation cross sections (see below). At present, there are no reasons to consider the fine structure in more detail.

The terms R_{nlj}^{+} and R_{nlj}^{-} in the balance equation (2.1.3) describe not only the secondary population and depopulation of a level with a given n, which can be neglected in a wide range of experimental conditions [9, 16], as cascade transitions, stepwise excitation and ionization, dissociative recombination, electron-ion recombination[1], quenching collisions, resonant radiation trapping, etc., which are neglected in this work. These terms describe also the processes redistributing the population between sublevels of the fine structure, with no change of n. The energy difference of the fine structure terms are as small as 0.1 cm^{-1} $\approx$ 0.14 K. The term energies are given in [21]. Hence, redistribution of the population can be caused by elastic

---

[1] Sometimes e+H^{+} recombination is included into collisional-radiative models. In the case of gas discharge plasmas it is meaningless, because the concentration of H$_3^{+}$ is usually higher than that of H^{+} and H$_2^{+}$ ions.

collisions of excited atoms with molecules, atoms, ions, and electrons, as well as by interaction with electromagnetic fields. It is well known that metastable $2^2S_{1/2}$ and resonance $2^2P_{1/2}$, $2^2P_{3/2}$ sublevels are usually mixed in plasmas. The processes (2.1.1) and (2.1.2) populate the nl- fine structure sublevels with different probabilities. In addition, the nl- sublevels have different radiative life times and transition probabilities. Therefore, the redistribution of the population densities between the fine structure levels, without changing n, influences the emission of Balmer lines. Currently, the data on the rates of these processes are insufficient for a strict consideration or numerical simulation. It is impossible to take into account such "mixing" processes in a general form; therefore, we analyze two simple limit *cases*:

> *Case 1*, when the mixing of the population densities of the sublevels is extremely small and can be neglected. This situation is typical for gas-beam and crossed beam experiments to determine collision cross sections,
>
> and
>
> *Case 2*, when the mixing processes are dominant for the formation of the population density distribution among various fine structure sublevels. This situation could appear in plasmas with high enough densities of particles and fields.

In both limit cases the total intensity of Balmer lines corresponding to a n →2 transition, the magnitude usually measured in experiments, may be presented in the same form

$$I_{n\to 2} = \sum I_{2\overline{1j}}^{nlj} = [H]n_e K_{dir}^{em}(F(\varepsilon)|n \to 2) + [H_2]n_e K_{dis}^{em}(F(\varepsilon)|n \to 2), \qquad (2.1.8)$$

where summation is carried out over all the components of the fine structure taking into account the selection rules for electric dipole transitions. $K^{em}(F(\varepsilon)|n \to 2)$ are the so-called line excitation rate coefficients or emission rate coefficients related to the corresponding cross sections $\sigma^{em}(\varepsilon)$ of the line excitation or emission cross section. Connection between those emission rate coefficients and emission cross sections is described by equations similar to (2.1.4).

But the meanings of explicit expressions for the emission rate coefficients, $K_{dir}^{em}(F(\varepsilon)|n \to 2)$ – for direct excitation and $K_{dis}^{em}(F(\varepsilon)|n \to 2)$ – for dissociative excitation by electron impact, in those limit *cases* are significantly different.

In *case 1*, in the balance equation (2.1.3) the terms R_{nlj}^+ and R_{nlj}^- are neglected and, using (2.1.6), the balance equation for all the fine structure components could be obtained. Then, the summation of the component intensities and by taking into account (2.1.7), lead to the total intensity of the Balmer series line $(n \rightarrow 2)$ in the form of (2.1.8) with the excitation rates written as

$$K_{dir}^{em1}\left(F(\varepsilon)| n \rightarrow 2\right) = \sum_{l,\tilde{l}} A_{2\tilde{l}}^{nl}\, \tau_{nl}\, K_{dir}^{ex}\left(F(\varepsilon)| nl \leftarrow 1s\right), \qquad (2.1.9)$$

$$K_{dis}^{em1}\left(F(\varepsilon)| n \rightarrow 2\right) = \sum_{l,\tilde{l}} A_{2\tilde{l}}^{nl}\, \tau_{nl}\, K_{dis}^{ex}\left(F(\varepsilon)| nl \leftarrow X\right). \qquad (2.1.10)$$

The upper index '1' in the emission rate coefficients points out the *case 1*; $A_{2\tilde{l}}^{nl}$, τ_{nl} - radiative transition probabilities and life times of the nl-sublevels. The excitation rate coefficients of the nl-sublevels are:

$$K_{dir}^{ex}\left(F(\varepsilon)| nl \leftarrow 1s\right) = \sum_{j} K_{dir}^{ex}\left(F(\varepsilon)| nlj \leftarrow 1s\right), \qquad (2.1.11)$$

$$K_{dis}^{ex}\left(F(\varepsilon)| nl \leftarrow X\right) = \sum_{j} K_{dis}^{ev}\left(F(\varepsilon)| nlj \leftarrow X\right), \qquad (2.1.12)$$

$K_{dir}^{ex}\left(F(\varepsilon)| nlj \leftarrow 1s\right)$ and $K_{dis}^{ex}\left(F(\varepsilon)| nlj \leftarrow X\right)$ are excitation rate coefficients of separate nlj-fine structure sublevel for direct and dissociative electron impact excitation.

In the *case 1* the atoms decay from the states, where they have been excited and the emission rate coefficient of the Balmer line is the sum of emission rate coefficients of separated fine structure components.

It is obvious that identical relations are also valid for the cross sections of the line excitation at any electron energy ε,

$$\sigma_{dir}^{em1}\left(\varepsilon| n \rightarrow 2\right) = \sum_{l,\tilde{l}} A_{2\tilde{l}}^{nl}\, \tau_{nl}\, \sigma_{nl \leftarrow 1s}^{ex}(\varepsilon), \qquad (2.1.13)$$

$$\sigma_{dis}^{em1}\left(\varepsilon| n \rightarrow 2\right) = \sum_{l,\tilde{l}} A_{2\tilde{l}}^{nl}\, \tau_{nl}\, \sigma_{nl \leftarrow X}^{ex}(\varepsilon). \qquad (2.1.14)$$

It is evident that the line excitation rate coefficients (2.1.9) and (2.1.10) are in complicated manner related to the partial excitation cross sections (2.1.13) and (2.1.14) of the fine structure sublevels.

If the cross sections are measured by the total intensities of the Balmer lines, then the partial cross sections of the nl-sublevel excitation cannot be determined. Even it cannot be done for the total cross section of a level with a given n, which is the sum of partial cross sections over l.

In the limit *case 2*, which is much closer to plasma conditions, the frequency of the above-mentioned processes leading to the redistribution of populations is much higher than the frequency of spontaneous emission from any sublevel of the fine structure. In this case, the distribution of populations should be close to the Boltzmann one. Taking into account that the difference of sublevel energies is negligible, leads to

$$N_{nlj} = N_n \frac{g_{nlj}}{\sum_{l,j} g_{nlj}},$$ (2.1.15)

where N_n is the sum of populations of all the fine structure nlj-sublevels of the nth level and g_{nlj} is the statistical weight of the nlj-sublevel.

Substituting (2.1.15) in (2.1.3) and (2.1.6) and summing over the fine structure components, within the above assumptions, the following expression is found for the rate coefficient of spectral line excitation due to direct and dissociative excitation

$$K_{dir}^{em2}\left(F(\varepsilon)\mid n \to 2\right) = A_{n\to2}\tau_n K_{dir}^{ex}\left(F(\varepsilon)\mid n \leftarrow 1s\right),$$ (2.1.16)

$$K_{dis}^{em2}\left(F(\varepsilon)\mid n \to 2\right) = A_{n\to2}\tau_n K_{dis}^{ex}\left(F(\varepsilon)\mid n \leftarrow X\right),$$ (2.1.17)

where the upper index '2' in emission rate coefficients points out the *case 2*. The effective radiative transition probabilities of the spontaneous emission of the $n \to \tilde{n}$ line are

$$A_{n\to\tilde{n}} = \frac{\sum_{l,l} g_{nl} A_{\tilde{n}l}^{nl}}{\sum_{l} g_{nl}},$$ (2.1.18),

the effective radiative lifetimes of the n-level are

$$\tau_n = \left(\sum_l \frac{g_{nl}}{\tau_{nl}\sum_l g_{nl}}\right)^{-1} = \left(\sum_{\tilde{n}=1}^{n-1} A_{n\to\tilde{n}}\right)^{-1}$$ (2.1.19),

and the total excitation rate coefficients

$$K_{dir}^{ex}\left(F(\varepsilon)\mid n \leftarrow 1\right) = \sum_l K_{dir}^{ex}\left(F(\varepsilon)\mid nl \leftarrow 1s\right) = \sum_l \sum_j K_{dir}^{ex}\left(F(\varepsilon)\mid nlj \leftarrow 1s\right),$$ (2.1.20)

$$K_{dis}^{ex}\left(F(\varepsilon)\mid n \leftarrow X\right) = \sum_l K_{dis}^{ex}\left(F(\varepsilon)\mid nl \leftarrow X\right) = \sum_l \sum_j K_{dis}^{ex}\left(F(\varepsilon)\mid nlj \leftarrow X\right).$$ (2.1.21)

It is necessary to emphasize, that the effective radiative probability eq. (2.1.18) and the lifetime eq. (2.1.19) have a physical meaning only in the case when the population distribution among the fine structure sublevels is in accordance with Bolzmann`s law.

In the limit *case 2* the emitting sublevel of the fine structure and the sublevel, which was excited, are independent. The emission rate coefficients of the Balmer line transition $n \to 2$ are proportional to the total excitation rate coefficient of the n-level as well as to its effective radiative probability and lifetime.

It is evident that relations analogous to (2.1.16), (2.1.17) and (2.1.20), (2.1.21) are also valid for corresponding cross sections, at any energy ε. Therefore, in considering *case 2*, certain effective excitation cross sections of spectral lines can be introduced:

$$\sigma_{dir}^{em2}(\varepsilon \,|\, n \to 2) = A_{n \to 2} \tau_n \sum_l \sigma_{nl \leftarrow 1s}^{ex}(\varepsilon), \qquad (2.1.22)$$

$$\sigma_{dis}^{em2}(\varepsilon \,|\, n \to 2) = A_{n \to 2} \tau_n \sum_l \sigma_{nl \leftarrow X}^{ex}(\varepsilon), \qquad (2.1.23)$$

which should be used to calculate the rate coefficients (2.1.16) and (2.1.17).

In this *case 2*, the effective cross section $\sigma^{em2}(\varepsilon \,|\, n \to 2)$ appears proportional to the total cross section of the nth level excitation, which is equal to the sum of all the partial cross sections $\sigma_{nlj \leftarrow 1s}^{ex}(\varepsilon)$ or $\sigma_{nlj \leftarrow X}^{ex}(\varepsilon)$. Thus, to calculate effective cross sections (2.1.22) and (2.1.23), one should know the partial cross sections as function of incident electron energies. These values can be found either by *ab initio* calculations or by special experiments with selective detection of populations of separate nl-sublevels.

2.1.2. Excitation of the Fulcher-α (2-2)Q1 line of molecular hydrogen

As well as for atomic lines, a simple excitation-deactivation model only including electron impact excitation and spontaneous emission is considered to describe the emission of a selected single molecular line, (2-2)Q1 of the Fulcher-α band system.

First details of the electron impact excitation of the rovibronic molecular level $d^3\Pi_u^-$, $v = 2$, $N = 1$ from the electronic ground state should be discussed here. In the transitions $d^3\Pi_u^-$, v, N $\leftarrow$ X$^1\Sigma_g^+$, v', N' the change of the rotational quantum number $\Delta N \equiv N\text{-}N'$ can only be even ($\Delta N = 0, \pm 2, \pm 4\ldots$) [23]; because of the extremely weak probability of a change in the total nuclei spin, even in collisions of molecules. Ortho- and para-hydrogen behave as different gases [24]. Neglecting transitions with changing angular momentum $\Delta N \geq 2$ leads to great simplifications of the analysis of the electron impact excitation, discussed in [25-27]. Therefore, in the case, when the concentration of vibronically excited molecules can be neglected, the excitation of the $d^3\Pi_u^-$, $v = 2$, $N = 1$ level occurs only from the $X^1\Sigma_g^+$ $v = 0$, $N=1$ level. In this case the expression for the intensity of the (2-2)Q1 line can be written:

$$I_{22Q1} = [H_2]_{X01} n_e A_{a21}^{d^-21} \tau_{d^-21} K_{mol}^{ex} (F(\varepsilon)| d^-21 \leftarrow X01), \qquad (2.1.24)$$

where $[H_2]_{X01}$ is the molecule concentration in the $d^3\Pi_u^-$, $v = 2$, $N = 1$ state, τ_{d^-21} is the radiative life time of the $d^3\Pi_u^-$, $v = 2$, $N = 1$ state [28]. $A_{a21}^{d^-21}$ is the probability of the $d^3\Pi_u^-$, $v = 2$, $N = 1 \rightarrow a^3\Sigma_g^+$ $v = 2$, $N = 1$ radiative transition [7] and $K_{mol}^{ex} (F(\varepsilon)| d^-21 \leftarrow X01)$ is the excitation rate coefficient for the electron impact of the $d^3\Pi_u^-$, $v = 2$, $N = 1$ state.

In order to express the intensity via the total concentration of molecules $[H_2]$ the ratio $[H_2]_{X01}/[H_2]$ has to be considered. In a gas at thermodynamic equilibrium, the ratio of ortho- and para-hydrogen is in accordance with Boltzmann's law and is a certain function of the temperature (T_0) [29]:

$$x_0(T_0) \equiv \frac{[H_2]_{orto}}{[H_2]_{para}} = 3 \frac{\sum_{N=1,3,5...} (2N+1) \exp\left(-\frac{E_N}{kT_0}\right)}{\sum_{N=0,2,4...} (2N+1) \exp\left(-\frac{E_N}{kT_0}\right)}, \qquad (2.1.25)$$

where $[H_2]_{ortho}$ and $[H_2]_{para}$ are the concentrations of ortho- and para-hydrogen respectively and E_N is the energy of the rotational levels of the ground state.

Heating of the gas, which is associated with the flow of the discharge current, or forced cooling, e.g. via walls of the discharge volume cooled by liquid nitrogen occurs only for a short time during the experiment [23], while the gas is stored at the room temperature for a long time. The time of an experiment is usually not long enough to change noticeably the ratio of ortho- and para-hydrogen. So if processes, which lead to a change of total nuclei spin, are neglected, then the ratio of ortho- and paramolecules remains the same determined by the room temperature T_0. Since the population distribution over rotational levels of the ground state is closely connected with the collisions of the molecules in plasma, it should be described by the Boltzmann law with the gas temperature T_g. For ortho-hydrogen follows:

$$[H_2]_{ortho} = [H_2]_{X01} \sum_{N=1,3,5...} \frac{2N+1}{3} \exp\left(-\frac{E_N - E_1}{kT_g}\right). \qquad (2.1.26)$$

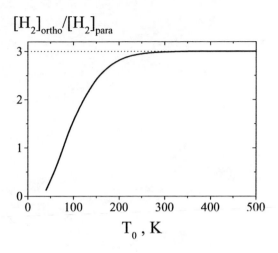

Fig. 2.1.1 The concentration ratio of ortho- and para-hydrogen $[H_2]_{ortho}/[H_2]_{para}$ as function of the temperature T_0 at the thermodynamic equilibrium.

The total concentration of hydrogen is the sum of the concentrations of ortho- and para-molecules:

$$[H_2] = [H_2]_{ortho} + [H_2]_{para} \qquad (2.1.27)$$

Taking into account (2.1.25), (2.1.26) and eq.(2.1.27), the ratio $[H_2]_{X01}/[H_2]$ can be expressed as follows:

$$\frac{[H_2]_{X01}}{[H_2]} \equiv \eta(T_g, T_0) = \frac{3x_0(T_0)}{(1 + x_0(T_0)) \sum\limits_{N=1,3,5...} (2N+1)\exp\left(-\dfrac{E_N - E_1}{kT_g}\right)}. \qquad (2.1.28)$$

The function $x_0(T_0)$ is shown on fig. 2.1.1. One may see that for room temperature and higher the concentration ratio of ortho- and para-hydrogen does not depend on T_0. Since gas is usually stored at room temperature this ratio is 3. Therefore the ratio $[H_2]_{X01}/[H_2]$ is only a function of the gas temperature T_g:

$$\eta(T_g) = \frac{9}{4} \frac{1}{\sum\limits_{N=1,3,5...} (2N+1)\exp\left(-\dfrac{E_N - E_1}{kT_g}\right)}. \qquad (2.1.29)$$

Taking into account the equations (2.1.28, 2.1.29), the intensity of the (2-2)Q1 line (2.1.24) may be described as:

$$I_{22Q1} = [H_2] \eta(T_g) n_e A_{u21}^{d^-21} \tau_{d^-21} K_{mol}^{ex} (F(\varepsilon) | d^-21 \leftarrow X01). \tag{2.1.30}$$

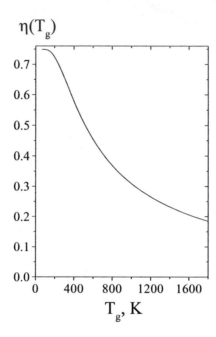

Fig. 2.1.2 The ratio of the population density of the first rotational level of the ground electronic-vibronic state and the total density of molecular hydrogen $\eta(T_g)$ as function of the gas temperature T_g.

It is possible to see that the intensity of the separate rovibronic molecular line depends not only on the concentration of molecules in the ground state but also on the gas temperature, as well as from the EEDF, electron concentration and molecular properties. In fig. 2.1.2 the function $\eta(T_g)$ is given. The function $\eta(T_g)$ depends rather strong on the gas temperature, reflecting the dependence of the (2-2)Q1 line intensity on T_g, eq.(2.1.30). So a temperature change from 300 K to 1000 K leads to changes of the intensity of the (2-2)Q1 line by a factor of more than two (fig. 2.1.2). Therefore, for the correct interpretation of the molecular line intensity the gas temperature is necessary to be known.

2.2. Analysis of the data about cross sections and rates coefficients

2.2.1 Data about cross sections of the hydrogen atom

All the known experimental data on the excitation cross sections of H_α and H_β lines are shown in fig. 2.2.1. The authors of experimental works tried to achieve conditions close to limit *case 1* considered above. As indicated in [30], the difference of the excitation function of [31] from others is probably caused by the population redistribution between the fine structure sublevels under an electric field. It was also indicated that the data of [32], normalized to a calculation in the Born approximation at high energies, are about 15% lower than the data of [30].

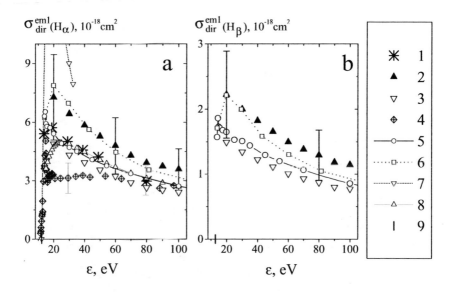

Fig.2.2.1 Cross sections ($10^{-18}cm^2$) of direct excitation of the H_α (a) and H_β (b) lines by electron impact as a function of the incident electron energy. The experimental data of (1) 1976 [30], (2) 1974 [32], (3) [32] normalized to the Born approximation [39] at 200 eV, and (4) 1968 [31]. The data calculated by the (5) convergent close-coupling method, 1999 [38]; (6) Born approximation, 1965 [22]; (7) R-matrix approximation, 1994 [36], maximum is not shown due to high overestimation in comparison to the other data; and (8) eikonal approximation, 1990 [37]; (9) indicate a threshold of the considered process.

One can see from fig. 2.2.1 that the results of [30], within experimental errors, coincide with the absolute measurements of [32], and these works have similar excitation functions. The data of [30] have the smallest experimental error; therefore, these are considered as most reliable. It is necessary to note that [30] is a single paper, where the partial cross sections of direct excitation of separate 3s-, 3p-, and 3d-sublevels of the fine structure were measured. For other lines of the Balmer series (Hβ, Hγ, Hδ, and Hε), only the emission cross sections of direct excitation of lines have been measured in a single work [32].

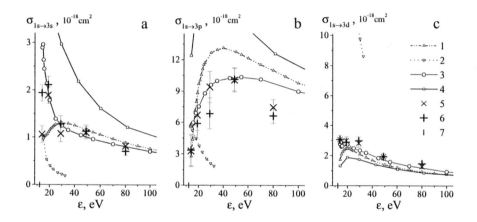

Fig. 2.2.2 Cross sections (10^{-18}cm^2) of hydrogen atom 3s- (a), 3p- (b), and 3d- sublevels (c) excitation by electron impact. The data calculated by the (1) eikonal approximation, 1990 [37]; (2) R-matrix approximation, 1994 [36]; (3) convergent close-coupling method, 1999 [38]; and (4) Born approximation, 1965 [39]; (5) and (6) are experimental data, 1976 [30]; (7) indicate a threshold of the considered process.

Theoretical calculations of the excitation cross sections for direct electron impact have been carried out separately for electron energies close to and much higher than the threshold values. The cross sections in the subthreshold region, see the review in [33], which are calculated in a very narrow energy range, lower than 1 eV, are characterized by an complicate resonant structure. This significantly complicates the involvement of these data in calculations of the reaction rate coefficients for applied problems. Currently, there are no experimental data in this energy range for exciting levels with n ≥ 3 at all, because one fails to achieve the necessary monokinetics of an electron beam. Therefore, a linear dependence of the cross sections from the threshold to the first measured or calculated point for all the processes is assumed.

Calculations at high electron energies have been carried out more frequently (see, e.g., [34, 35]). Figure 2.2.2 displays the calculated cross sections of direct excitation of $3s$-, $3p$-, and $3d$-sublevels published in recent papers [36–38] and the experimental values of [30]. It is evident from figs. 2.2.1 and 2.2.2 that the data, obtained by the convergent close-coupling (CCC) method in [38], fit much better both the total cross section of the H_α and H_β line excitation and the partial cross sections of $3s$-, $3p$-, and $3d$-sublevel excitation of the fine structure than other data in the literature. Furthermore, the agreement of the calculation of [38] with the experimental data can be considered as satisfactory taking into account that

(i) the excitation function of the $3s$-state sharply grows in the sub-threshold range, first two experimental points of [30] are at 13.8 eV and the first calculated point of [38] is at 14 eV;

(ii) a slight systematic overestimation of the experimental cross sections of the $3d$-sublevel excitation (see fig. 2.2.2c) is probably due to the fact that the authors of [30] could not take into account the cascade transitions from the $4f$- and $5f$-sublevels because of the absence of Born cross sections for these sublevels;

(iii) the reason of the disagreement between the experimental and calculated data for the $3p$-sublevel at 80 eV, see fig. 2.2.2b, remains unclear; however, this does not have a strong effect on the calculation of the cross section of the H_α line excitation, where the data coincide within experimental errors (see Fig. 2.2.1a). We may also expect that the EEDF steeply slopes in the electron energy range above 50 eV. (Based on the assumption of the Maxwellian EEDF, this cross section component begins to contribute about 1% to the reaction rate at an electron temperature of 7 eV.)

Additional investigations are necessary to explain the difference between the calculated and experimental cross sections of the excitation of the $3p$-sublevel. The method of the measurement of partial cross sections [30] is based on the difference between radiative lifetimes of sublevels. The lifetimes of $3s$- and $3d$-sublevels differ by an order of magnitude, while those of $3d$- and $3p$- sublevels differ only threefold. According to ref. [30], the separation of a signal related to the $3p$-component is rather complicate. This can also be concluded from the fact that the difference between the calculated [38] and experimental data at 80 eV is

23

comparable to the difference between the experimental points at 29 eV. Unfortunately at present, it is difficult to come to a more certain result.

The CCC calculations, presented in ref. [38], were also carried out for the partial cross sections of the excitation of the nl-sublevel ($n = 4$). The results are in well coincidence with the cross section of the H_β line excitation measured in [32] (see fig. 2.2.1b). The data of [38] are the only complete results for the $n = 4$ level. (In [39], the cross section of f-sublevels was not reported.) In the following, the considerations will be based on the cross sections of [38] for hydrogen atom excitation by direct electron impact.

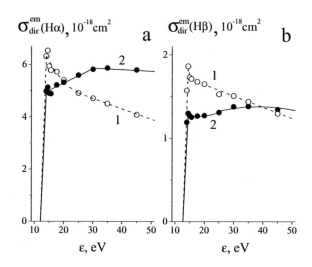

Fig. 2.2.3 Emission cross sections of H_α (a) and H_β (b) line excitation by direct electron impact, calculated for the two limit *cases* by formulas (2.1.13) (*case 1*) and (2.1.22) (*case 2*). Dashed and solid lines show the results of sectional polynomial approximations used in this study.

Based on the data of ref. [38], the emission cross sections of the H_α line for the limit *cases 1* and 2 have been calculated by formulas (2.1.13) and (2.1.21) for few electron energies given in [38]. To calculate the rate coefficients, those few points were sectionally approximated by polynomials. The results are given in appendix, table 3, and are shown in fig. 2.2.3. One can see, that redistributions of populations of the fine structure sublevels change not only the absolute value of the cross sections of line excitation, but also the excitation function as a function of the electron energy.

Unfortunately, only poor information on the dissociative excitation of the lines of the Balmer series is available at present. The emission cross sections of the dissociative excitation of the Balmer lines were measured in several works. These works have been reviewed in detail by Möhlmann et.al in ref. [40]. The function of the dissociative excitation of the H_α and H_β lines, measured in [41], coincides well with the data of [40]. Note that monochromators were used for radiation detection only in [40] and [41]. In other works light filters were employed. The problem of extraction of the disturbing radiation complicates considerably the experimental technique; therefore, the data of [40] are considered as most reliable. The data of [41] are cited in [40] as a private communication.

The partial cross sections of the dissociative excitation of the nl-sublevel for $n = 3$ and 4 were measured only in [42–44]. The authors of [42] measured the ratios of the partial cross sections of dissociative excitation of the $3s$-, $3p$-, and $3d$-sublevels to the emission cross section of the H_α line at electron energies of 1000 and 2000 eV. In [43], a similar ratio was measured for the cross section of the excitation to the $3s$-level at 300 eV. The corresponding ratios measured at different energies were found to coincide. Therefore, here it is assumed that these ratios are independent on energy. This allows it to determine the partial cross sections of the dissociative excitation of $3s$-, $3p$-, and $3d$-sublevels. Using the excitation function measured in [40] the emission cross section of the dissociative excitation of the H_α line assuming a Boltzmann distribution of the populations corresponding to *case 2* have been calculated. The data of [41] were published in less detail. This makes it possible to estimate only the quantitative difference between the cross section measured in the beam experiment and that corresponding to the plasma conditions. At present, it seems to be impossible to determine the difference in the shapes of the cross section because of a lack of experimental data.

The partial cross sections of the excitation of nl-sublevels ($n = 4$) were measured only in [44] and only for five incident electron energies: 19, 24, 32, 50, and 100 eV. The cross section of dissociative excitation has a complicate shape [40]; the interpolation of the cross section between the five known points leads to a distinct level of uncertainty. The cross sections of dissociative excitation of *case 2* between the five points, found by formula (2.1.23), were interpolated so that the excitation function is close to that of the dissociative excitation of the H_β line of *case 1*, measured in the beam experiment [40]. The analysis of the difference of the cross sections calculated in the two limit *cases 1* and *2* needs obviously more detailed experimental data on the dependencies of the partial cross sections on the collision energy.

The considered data were used to approximate the cross sections of dissociative excitation of the H_α and H_β lines for *cases 1* and *2* by sectional polynomial functions. The results of approximations of both lines are listed in the appendix, table 3. It is evident that the difference between the cross sections of dissociative excitation for the two limit *cases* is much larger than the same difference for excitation by direct electron impact. In particular, the cross sections of dissociative excitation of the H_β line, calculated by (2.1.14) and (2.1.23), differ by one order of magnitude.

The cross sections of dissociative excitation of the H_α and H_β lines were first found in this study assuming a Boltzmann distribution of population over the fine structure sublevels (*case 2*). These values are based on a number of assumptions because of a lack of experimental data. In refs. [9 - 12, 16] in order to take into account the dissociative excitation in plasma, the cross sections measured in beam experiments were used, which correspond to *case 1* of the present study. The excitation cross sections found in the limit *case 2* are more consistent with the plasma conditions. Hence, the significant difference between the emission cross sections of the Balmer series in *cases 1* and *2* leads to doubts on the precision of numerical results of previous works, related to the determination of the degree of dissociation of hydrogen in plasma.

The lack of data on the cross sections of the excitation of *nl*-sublevel by electron impact allows a detailed analysis only of the direct excitation levels of the hydrogen atom with n ≤ 4. The use of the Born cross sections [22], [39] may result in systematic errors, since the Born approximation is known to yield in systematically overestimated cross sections, see fig. 2.2.1. The cross sections of the dissociative excitation of the H_α and H_β lines can be quantitatively estimated. Due to a great amount of "mixing" in the emission of these lines, even a rough account of the fine structure for the dissociative excitation, proposed in the present study, is rather important.

2.2.2 Data about rate coefficients of the hydrogen atom

To analyze the result of taking the fine structure into account in the excitation and deactivation kinetics, the reaction rate coefficients have been calculated with a Maxwell EEDF by formula (2.1.5), using the effective cross sections of line excitation given in this study (see table 3). The calculated data are listed in the appendix, table 4.

To visualize the data, fig. 2.2.4 shows the ratios of the rates corresponding to *cases 1* and *2*. One can see the significant difference between the rates corresponding to the conditions of the beam (*case 1*) and plasma (*case 2*) experiments. The ratio of these rates depends on the electron temperature (hence, on the shape of the EEDF).

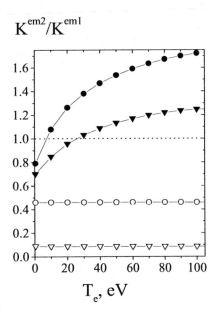

Fig. 2.2.4 The ratio of emission rate coefficients found in the two limit cases K^{em2}/K^{em1}, for direct excitation of H_α ($\bullet$) and H_β ($\blacktriangledown$) and for dissociative excitation of H_α ($\bigcirc$) and H_β ($\bigtriangledown$) in dependence on the electron temperature T_e.

The rate coefficients of direct excitation of the H_α line, obtained in this study for plasma conditions (*case 2*), are compared to similar data from previous works in fig. 2.2.5. One can see that the rate data of various works poorly agree with each other.

The data of [45] are based on the Born cross sections, however, they were multiplied by a correction factor to compensate an overestimation of the Born cross section in the near-threshold region in contrast to the experimental data. These rate coefficients appeared to be much lower than those determined in this study, which indicates significant problems in the use of the Born cross sections.

The data of [46] are based on the cross sections calculated in the R-matrix approximation. Unfortunately, the cross section data were not published by the authors of [46], while the known recent calculations in ref. [36] by the R-matrix method yield a highly overestimated result in comparison to the experimental data of [30], see figs. 2.2.1 and 2.2.2. These rate coefficients exceed considerably those found in this study.

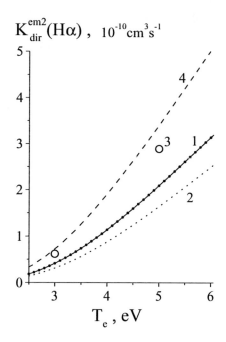

Fig. 2.2.5 Rate coefficients of direct excitation of H_α line by electron impact, corresponding to case 2: (1) the data of this study, (2) 1971 [45], (3) 2000 [46], and (4) Atomic Data and Analysis Structure, ADAS database [47].

The rates given in ref. [46] were incorporated in the ADAS, Atomic Data and Analysis Structure, database [47] and used in [48]. One of the authors, U. Fantz, has kindly made available numerical data on the rate coefficients. The data of [46] and of [48] are different. This is unclear, since the others use the same ADAS databases.

When calculating the H_α direct excitation rate coefficients, the cross sections of H_α line excitation of table 3 in the appendix were used. These cross sections are based on the data of [38] and coincide well with the only complete experimental data set of the partial cross sections of the excitation of the nl-sublevel of [30], figs. 2.2.1 and 2.2.2. Hence, based on the numerical difference between the rate coefficients of this study and [46], [47] it can be concluded that the cross sections used in [46], [47] coincide poorly with the experimental data of [30], hence, they are much less reliable.

2.2.3 Data about cross sections and rate coefficients of the hydrogen molecule

2.2.3a Introduction

Spectral lines of the Fulcher-α band system, the $d^3\Pi_u$ v, N $\rightarrow$ a$^3\Sigma_g$' v', N' radiative transitions, of molecular hydrogen, have special attraction for plasma diagnostics. These spectral lines have relatively strong intensities in a wide range of experimental conditions and they are situated in a comfortable visible part of the spectrum. The cross sections of the excitation of the $d^3\Pi_u$ state have been studied in a number of experimental [49-52], and theoretical [53-56] works. But numerical data about excitation cross sections of single ro-vibronic levels of the $d^3\Pi_u^-$ state have not been published so far. The main problem to analyze the data available in literature is the comparison of theoretical data on excitation cross sections of the $d^3\Pi_u$ state and experimental data on emission cross sections of spectral lines of the Fulcher-α band system. The relation between excitation and emission cross sections obtained in the works [49-56] is not obvious. In the following it is described how the available data of the cross sections have been analyzed and how the numerical data for *separate ro-vibronic transition* have been obtained.

General quantum mechanic analysis of the excitation of molecular levels is very complicated [53]. For this reason in theoretical studies often the ro-vibronical structure of the molecular terms is neglected, as in ref. [54, 55]. In ref. [56] the excitation cross sections of electronic-vibronic levels of the $d^3\Pi_u$ state were calculated without separation of the rotational structure. How the excitation cross sections of single ro-vibronical levels could be obtained from data of [56] remains unclear. In refs. [52, 53] the semi-empirical way of determination of the excitation cross sections of single ro-vibronical level have been developed. But the authors were based on erroneous assumptions of deviations from the Franck-Condon approximation due to erroneous information about radiative transition probabilities and life times, which was corrected later, (see [7]). Therefore, in summary no complete theoretical data of cross sections are available.

2.2.3b Relation between the excitation functions of the $d^3\Pi_u$ state and $d^3\Pi_u^-$ v, N level

From a general consideration of the Franck-Condon approximation follows that the excitation function should not depend on vibronic or rotational quantum numbers [53], at least in some ranges of the electron energy. It is necessary to note, that it cannot be valid at all electron energies, because the cross sections of different ro-vibronical transitions have different threshold energies. In ref. [56]

(i)   the excitation cross sections of the transitions $X^1\Sigma_g$ v=0 $\rightarrow$ $d^3\Pi_u$ v' = 0, 1, 2, 3;

(ii)   the sum of cross sections of the transitions $X^1\Sigma_g$ v=0 $\rightarrow$ $d^3\Pi_u$ v' = 0, 1, 2, 3 and

(iii)   the total cross section of the $d^3\Pi_u$ state, unresolved over vibronic levels

are given.

All these cross sections normalized at 20eV are shown on fig. 2.2.6. One may see that for electron energies higher than 20 eV all presented excitation functions are almost the same, i.e. they do not depend on the vibrational transition. It cad be assumed that the excitation functions of separate ro-vibronic levels are not too different from the excitation function of the electronic level calculated without taking into account the ro-vibronic structure.

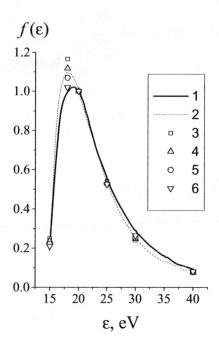

Fig. 2.2.6 Excitation functions the of $d^3\Pi_u$ state from [56] normalized at 20 eV. (1) is the total cross section, (2) is the sum of the cross sections for first four vibronic levels, (3),(4), (5), (6) are the cross sections of the excitation of levels with v=0, 1, 2, 3 correspondingly.

The experimental studies of refs. [49], [50] are devoted to the determination of the emission cross section of the total intensities of the Fulcher-α system. The emission of the Q branch of the Fulcher-α system comes from the $d^3\Pi_u^-$ state while the P and R branches radiates from $d^3\Pi_u^+$ state, see fig.2.2.7, [23]. The $d^3\Pi_u^+$ state is perturbed by levels of the $e^3\Sigma_u^+$ state [50], [57] and achieving information about excitation cross sections of levels from P and R branches is rather difficult. So in refs. [49] and [50] the excitation function of the $d^3\Pi_u^-$ state has been measured from a sum of the intensities of the diagonal Fulcher bands, i.e. from the Q branch, $d^3\Pi_u^- \rightarrow a^3\Sigma_g^+$ v=0, 1, 2, 3.

Unfortunately the authors of [49], [50] did not specify how they make the summation of the intensities, which are belongs to the same vibrational transition. In ref. [49] the gas temperature was estimated to 350 K from the ratio of the intensities of lines with N = 1 and 3, i.e. from Q1 and Q3 lines. Normally these lines are the strongest in the rotational structure at low gas temperature. It can be assumed that only these lines were taken into account and the excitation function of the sum of the Q1 and Q3 line was measured. The authors report the total cross section of the whole Fulcher system, but the influence of the radiation of the P and R branches was taking into account by a calculated correction factor, which does not depends on electron energy. Therefore in ref. [49] and [50] the authors report the excitation functions of the diagonal Fulcher bands, or of the $d^3\Pi_u^-$ state, as excitation functions of whole Fulcher system, or of the $d^3\Pi_u$ state. This report leads to a certain confusion, because the excitation mechanisms of the $d^3\Pi_u^+$ and $d^3\Pi_u^-$ levels are different.

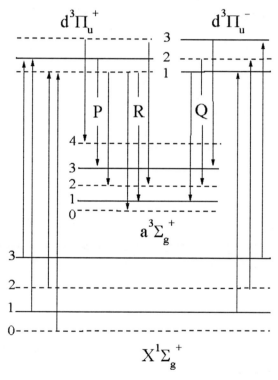

Fig. 2.2.7 Diagram of the excitation of the Fulcher-α band system. Solid and dashed lines show energy levels of ortho- and para-hydrogen correspondingly.

The rotational levels with the same N of the $d^3\Pi_u^+$ and $d^3\Pi_u^-$ states have a different symmetry of the nuclei, i.e. they belong ortho- and para-hydrogen [23]. The excitation of the $d^3\Pi_u^-$ levels occurs mostly in a transitions with $\Delta N = 0$, as it was described in chapter 2.1.2, while the excitation of the $d^3\Pi_u^+$ levels occurs mostly in transition with $\Delta N = \pm 1$. It is not obvious that the excitation functions of $d^3\Pi_u^+$ and $d^3\Pi_u^-$ levels are identical.

In ref. [51] the measurement of the excitation function of single ro-vibronic lines of Q bands are reported. It was found, that these excitation functions are within the experimental error, of about 10%, are the same. This result gives an opportunity to consider the excitation function measured in [49] and [50] as the excitation function of a single ro-vibronic line or a single ro-vibronic level of the $d^3\Pi_u^-$ state.

2.2.3c Compilation of excitation function of a ro-vibronic level of the $d^3\Pi_u^-$ state

In fig. 2.2.8 all measured and calculated excitation functions are normalized at 22.5 eV, the first data point in [50]. The excitation function of the $d^3\Pi_u^-$ levels from work [49] is most accurate, because it was measured at the smallest pressure compared to the other studies [50-52]. Therefore, influences of secondary processes, which disturb the excitation function at electron energies higher than 29.4 eV, were minimized [52]. It is important to note, that in ref. [49] for the detection of radiation a light filter was used. In these measurements the diagonal Fulcher-α band was overlapped by other H_2 emissions. The authors calculated the influence of background radiation on the measured intensities depending on the electron energy, 5% at 15.6 eV and 50% at 100eV. So, additional experimental investigations are necessary.

In [55] the authors reported that their cross section of the $d^3\Pi_u$ state, calculated by Schwinger multichannel method, and the cross section from [56], calculated by distorted wave approximation, do not agree, and the reason remains to be explained. From fig. 2.2.8 one may see, that excitation functions calculated in these works are close to each other. Unfortunately, the authors of [55] and [56] did not make a detailed comparison of their results with calculations based on Born approximation and they did not provide any data for electron

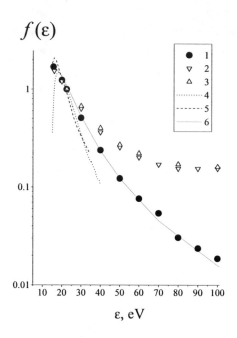

Fig.2.2.8 Excitation functions of the $d^3\Pi_u$ state of H_2 normalized at 22.5 eV. Based on the experimental studies: (1) [49], (2) [51, 52], (3) [50]; and on theoretical studies: (4) distorted wave approximation [56], (5) Schwinger multichannel method [55], (6) Born approximation [54].

energies higher than 40 eV. It is known that at high electron energies Born approximation is in good agreement with experimental data. For example, for atomic hydrogen at electron energies higher than 50 eV calculated and measured excitation functions have a good agreement and at 200 eV even absolute values of cross sections. So at present it is difficult to say something certain about the applicability of the data of ref. [55] and of ref. [56] for plasma diagnostics purposes and about the disagreement of these data with experiments and Born approximation [54].

From fig. 2.2.8 it can been seen, that the Born approximation [54] describes surprisingly well the experimental data of ref. [49]. Further, to obtain the excitation cross section of the $d^3\Pi_u^-$ v = 2, N = 1 level the measured excitation function of [49] are used, denoted as $f_{d \leftarrow X}(\varepsilon)$.

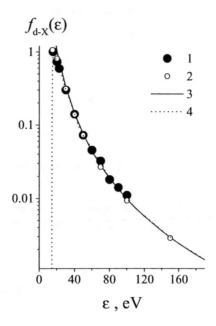

Fig. 2.2.9. Excitation function of $d^3\Pi_u^-$, level of the hydrogen molecule by electron impact, normalized on its maximum. (1) experimental data [49]; (2) Born approximation [54], normalized for fit of [49]; (3) present study, approximation by formula (2.1.31); (4) present study, sectional polynomial approximation.

For the extrapolation of the excitation function up to 1000 eV the eq. 2.1.31 was used, in accordance to Born approximation [58]:

$$f_{d \leftarrow x}(\varepsilon) = \frac{A}{(B+u)^3}$$ (2.1.31)

where A and B are adjusted parameters; u is, so-called in [22], the electron energy in "thresholds" units:

$$u = \frac{\varepsilon - 14.4}{14.4},$$

and ε is electron energy in eV. The experimental data of ref. [49] normalized on its maximum are plotted in fig. 2.2.9 by solid circles. The theoretical data of the Born approximation [54] were normalized to fit the data of [49] in particular at high electron energies, shown by open circles. The obtained set of points is fitted by formula (2.1.31). In fig. 2.2.9 the solid curve corresponds to the coefficients A=3.4 and B=1.1. For the data a sectional polynomial approximation has been applied, the results are shown on fig. 2.2.9, dotted curve, and in table 5 in the appendix.

2.2.3d Normalization of the cross section of the $d^3\Pi_u^-$ v=2, N=1 level

In gas- electron beam experiments, reported in [51], [52] the intensity ratio of the (2-2)Q1 line of the Fulcher-α system and of the H_α line has been measured at an electron energy of 50 eV, leading to $I_{22Q1}/I_{H\alpha}$=0.07 $\pm$ 0.03. This ratio has been used for a normalization of the excitation cross section of the $d^3\Pi_u^-$ v=2, N=1 level.

In the case of the molecule excitation by an electron beam, as in the experimental studies [51], [52] the expression for the intensity of the (2-2)Q1 line, eq. (2.1.30), at electron energy of 50 eV can be introduced in following way:

$$\frac{I_{22Q1}(50eV)}{A_{a21}^{d^-21}\tau_{d^-21}} = [H_2]\eta(T_g)\left(n_e\sqrt{\frac{2\varepsilon}{m_e}}\right)_{50eV}\sigma_{d^-21\leftarrow X01}^{max}f^{LOU}(50eV) \qquad (2.1.32)$$

where $\sigma_{d^-21\leftarrow X01}^{max}$ is a value of the excitation cross section of the $d^3\Pi_u^-,(v = 2, N = 1)$ level by electron impact from the ground state $X^1\Sigma_g^+$ v $=$ 0, N $=$ 1 in maximum; $f^{LOU}(\varepsilon)$ is the excitation function of the $d^3\Pi_u^-,(v=2, N=1)$ level at the experimental condition of refs. [51], [52] normalized on its maximum. n_e and m_e are the density and the mass of electrons. $A_{a21}^{d^-21}$ is the probability of the $d^3\Pi_u^-$ v=2, N=1 $\rightarrow$ a$^3\Sigma_g^+$ v=2, N=1 radiative transition [7], τ_{d^-21} is the radiative life time of the $d^3\Pi_u^-$ v=2, N=1 state [28], [H_2] is the concentration of hydrogen molecule in the electronic ground state, $\eta(T_g)$ is a function of the gas temperature, see eq. (2.1.29).

Taking into account that in a beam experiment the atomic density is low and the direct electron impact excitation is negligible, the expression for the intensity of the H_α line, eq. (2.1.8) can be written as:

$$I_{H\alpha}(50eV) = [H_2]\left(n_e\sqrt{\frac{2\varepsilon}{m_e}}\right)_{50eV} \sigma_{dis}^{eml}(50eV \mid 3\to 2). \qquad (2.1.32)$$

The division of the equations (2.1.31) and (2.1.32) leads to:

$$\sigma_{d^-21\leftarrow X01}^{max} = \frac{I_{22Q1}(50eV)}{I_{H\alpha}(50eV)}\frac{1}{\eta(T_g)}\sigma_{dis}^{eml}(50eV \mid 3\to 2)\frac{1}{f^{LOU}(50eV)A_{a21}^{d^-21}\tau_{d^-21}} \qquad (2.1.33)$$

Eq.(2.1.33) has been used to determine $\sigma_{d^-21\leftarrow X01}^{max}$. The gas temperatures in refs. [51], [52] were estimated as $345 \div 365$ K with the ratio $[H_2]_{ortho}/[H_2]_{para}=3$. This gives then $\eta(T_g)=0.61\pm0.01$, see eq. (2.1.29). $\tau_{d^-21} = (39.5\pm1.9)$ ns, see detailed discussions in [28], $A_{a21}^{d^-21} = (17\pm1)\cdot10^6$ s^{-1} was recommended in [7]. Assuming that the emission cross sections of H_α in ref. [52] and [40] are the same, both works are gas – beam experiments, than $\sigma_{dis}^{eml}(50eV \mid 3\to 2)=0.907\cdot10^{-18}$cm^2 measured in [40].

Substituting all these values in (2.1.33) leads to a value of the excitation cross section of

$$\sigma_{d^-21\leftarrow X01}^{max} = (0.9\pm0.4)\cdot10^{-18}cm^2.$$

Using this value and a sectional polynomial approximation of the excitation function, see table 5 of the appendix, the excitation rate coefficients, found with a Maxwellian EEDF, are listed in table 6 of the appendix.

2.3 Relation between ratio of the line intensities and plasma parameters

For a correct interpretation of the ratio of line intensities one needs to know the EEDF. Caused by known difficulties of Langmuir probe measurements of the EEDF in plasmas, the rate coefficients, calculated with a Maxwellian EEDF, are used for estimations in the present study (tables 4 and 6). The temperature of the Maxwellian approximation is treated as an effective parameter T_e^{eff}, which is not connected with the average electron energy in the plasma. But T_e^{eff} can be considered as a feature approximately describing the actual EEDF in the threshold region. In this case, the rate coefficients in eq. (2.1.8) and (2.1.30) become a function only of one parameter T_e^{eff}. Under this assumption the dependencies of the intensity ratios on plasma parameters: (i) the density ratio of atomic and molecular hydrogen [H]/[H₂], which characterize the degree of dissociation of hydrogen, see eq. 1.1 and (ii) parameter T_e^{eff}, which approximately describing the actual EEDF in the threshold region, are investigated in the following.

2.3.1 Intensity ratio of Balmer lines

In order to get a relationship between the intensity ratio of atomic lines, H$_\alpha$ ($I_{3\to2}$=$I_{H\alpha}$), H$_\beta$ ($I_{4\to2}$=$I_{H\beta}$), and plasma parameters, [H]/[H₂], T_e^{eff}, eq.(2.1.8) is used.

$$\frac{I_{3\to2}}{I_{4\to2}}=\frac{I_{H\alpha}}{I_{H\beta}}=\frac{\dfrac{[H]}{[H_2]}K_{dir}^{em}\left(T_e^{eff}\,|\,3\to2\right)+K_{dis}^{em}\left(T_e^{eff}\,|\,3\to2\right)}{\dfrac{[H]}{[H_2]}K_{dir}^{em}\left(T_e^{eff}\,|\,4\to2\right)+K_{dis}^{em}\left(T_e^{eff}\,|\,4\to2\right)}. \qquad (2.3.1)$$

One may see that, if (i) the concentration ratio [H]/[H₂] is very big and the dissociative excitation, eq (2.1.2), is negligible or (ii) [H]/[H₂] is very small and the direct excitation, (2.1.1), is negligible, then the intensity ratio is defined only by the ratio of the corresponding emission rate coefficients. In these cases no information about the degree of dissociation can be derived. Nevertheless, the intensity ratio could provide information about the parameter T_e^{eff}, which partly describes properties of the EEDF.

The relation (2.3.1) in dependence on the density ratio of atomic and molecular hydrogen and on the parameter T_e^{eff} is visualized in figs. 2.3.1 – 2.3.3. In fig. 2.3.1 the intensity ratio, calculated for T_e^{eff} = 5 eV for the two *limit cases*, and data from Schulz-von der Gathen and Doebele [12] are shown. One may see that the mixing processes change dramatically the relation between the intensity ratio of Balmer lines (H_α, H_β) and of [H]/[H₂].

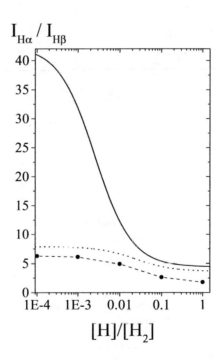

Fig. 2.3.1. The ratio of intensities of H_α and H_β lines $I_{H\alpha}/I_{H\beta}$ as function of the density ratio of atomic and molecular hydrogen, calculated at T_e^{eff} = 5eV. Dotted line- *case 1* (in accordance to beam experiments); solid line – *case 2* (in accordance to plasma experiments); (●) – data from [12].

In *case 2*, when mixing processes are dominant and fine structure sublevels populated in accordance to the Bolzmann law, the intensity ratio is much more sensitive to changes of the degree of dissociation than in *case 1*, when mixing processes are neglected. The results of [12], based on experimental emission cross sections of [32] and of [40], are close to the calculation of the present work for *case 1* (see fig. 2.3.1). It can be assumed, that the population distribution over the fine structure of levels is close to Bolzmann's law in a wide range of experimental conditions. Therefore, it seems to be more appropriate to use rate coefficients for *case 2* rather than those for *case 1*, used in [12].

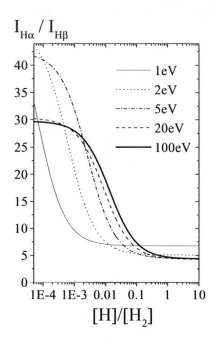

Fig 2.3.2. The ratio of intensities of H_α and H_β lines $I_{H\alpha}/I_{H\beta}$ as function of the density ratio of atomic and molecular hydrogen, calculated for *case 2* at various T_e^{eff}.

In fig. 2.3.2 the ratio of H_α and H_β intensities, eq. (2.3.1), as function of the density ratio of atomic and molecular hydrogen, calculated for *case 2* for various T_e^{eff}, is shown. It can be seen that:

(i) With increasing degree of dissociation the intensity ratio is decreasing, assuming the EEDF does not change its shape.

(ii). If $[H]/[H_2]$ is higher than 0.2 the intensity ratio is determined only by direct electron impact excitation, eq. (2.1.1), and does not depend on the degree of dissociation for any T_e^{eff}, and therefore, it can be assumed, for any shape of the EEDF. The lowest value of [H]/[H$_2$] strongly depends on T_e^{eff}, when direct excitation is negligible.

(iii) In the frame of the used excitation-deactivation model the ratio of H$_\alpha$ and H$_\beta$ intensities has a minimum at about 4.3. When the ratio is smaller as 5 then it can not reflect any changes of [H]/[H$_2$] or T_e^{eff}.

(iv) With increasing T_e^{eff} the sensitivity of the method on values of T_e^{eff} is decreasing, therefore uncertainties of the determination of the degree of dissociation related to an insufficient knowledge of the EEDF are decreasing. The curve corresponding to $T_e^{eff} = 100$ eV can be used for the determination of the upper limit of the degree of dissociation of hydrogen.

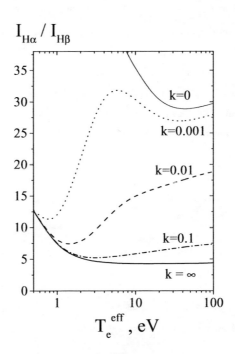

Fig. 2.3.3. The ratio of intensities of H$_\alpha$ and H$_\beta$ lines I$_{H\alpha}$/I$_{H\beta}$ as function of T_e^{eff} calculated for various ratios of densities of atomic and molecular hydrogen (k=[H]/[H$_2$]) for *case 2*.

The intensity ratio as function of the parameter T_e^{eff} calculated at various concentration ratios of atomic and molecular hydrogen is shown in fig. 2.3.3. One may see that:

(i) If the density of molecules is only 10 times bigger than the atomic density then the dissociative excitation could lead to noticeable changes of the intensity ratio, depending on the EEDF.

(ii) When both channels, eq. (2.1.1, 2.1.2), of excitation are involved, the intensity ratio increases with T_e^{eff}, caused by an increasing role of the dissociative excitation with T_e^{eff}.

(iii). When for the excitation of the line only direct electron impact is involved, the intensity ratio is sensitive to changes of T_e^{eff} up to $T_e^{\text{eff}} = 2$ eV. So, in plasmas with a high degree of dissociation ($[H]/[H_2] > 0.2$) the ratio of atomic lines can be used for the determination of parameters of the EEDF, but only if the EEDF rapidly drops between the excitation thresholds of the lines. Rate coefficients below $T_e^{\text{eff}} = 2$ eV strongly depend on the threshold behavior of the cross sections, which is unknown. Therefore, more detailed information about the threshold region of cross sections for the direct electron impact of H_α and H_β is required for progress in plasma diagnostic.

Thus, the intensity ratio of H_α and H_β lines can be used for the determination of the degree of dissociation of hydrogen in a wide range of experimental conditions; especially in plasmas with low degree of dissociation, when $[H]/[H_2]$ is below 0.2.

2.3.2. Intensity ratio of H_α and (2-2)Q1 lines

The ratio of equations (2.1.8) and (2.1.30) represents the intensity ratio of atomic and molecular lines as function of three plasma parameters, the density ratio $[H]/[H_2]$, the parameter T_e^{eff} and the gas temperature T_g:

$$\frac{I_{n\to 2}}{I_{22Q1}} = \frac{1}{\eta(T_g)} \frac{\frac{[H]}{[H_2]} K_{dir}^{em}\left(T_e^{\text{eff}} \mid n\to 2\right) + K_{dis}^{em}\left(T_e^{\text{eff}} \mid n\to 2\right)}{A_{a21}^{d^-21} \tau_{d^-21} K_{mol}^{ex}\left(T_e^{\text{eff}} \mid d^-21 \leftarrow X01\right)} \tag{2.3.2}$$

One may see that the intensity ratio is a linear function of [H]/[H$_2$]. When the atomic density is too small and direct excitation, eq. (2.1.1), can be neglected then the intensity ratio depends only on corresponding rate coefficients, which are functions of the parameter T_e^{eff}, and the gas temperature.

The relation (2.3.2) in dependence on the density ratio of atomic and molecular hydrogen and on the parameter T_e^{eff} is shown in figs. 2.3.4 – 2.3.6. The dependence on the gas temperature, presented by the function $\eta(T_g)$, is shown in fig. 2.1.2 and discussed in the previous chapter.

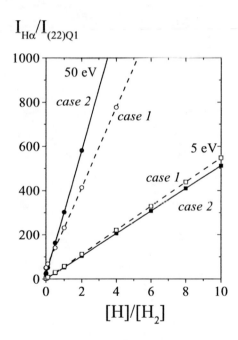

Fig. 2.3.4 The ratio of intensities of H$_\alpha$ and (2-2)Q1 lines I$_{H\alpha}$/I$_{(22)Q1}$ as function of the density ratio of atomic and molecular hydrogen [H]/[H$_2$], calculated at $T_g = 1000K$ and at $T_e^{eff} = 5$ eV and 50 eV. Dashed line - *case 1* (in accordance to a beam experiment); solid line – *case 2* (in accordance to a plasma experiment).

In fig. 2.3.4 the intensity ratio calculated for $T_e^{eff} = 5$ eV and for $T_e^{eff} = 50$ eV at the two limit cases is shown. The difference for the two limit cases is noticeable and depends on the parameter T_e^{eff}. The correct rate coefficients are important to use for the interpretation of the intensity ratio of the atomic and molecular lines. The dependencies in figs 2.3.5, 2.3.6 are calculated with rate coefficients for *case 2*, which is more suitable for plasma conditions.

In fig. 2.3.5 the intensity ratio is shown as function of the density ratio of atomic and molecular hydrogen at various T_e^{eff}. At $T_e^{eff} \geq 5eV$ for low values of the density ratio (≤ 0.08) the intensity ratio does not depend on changes of the degree of dissociation. At higher values of the density ratio the intensity ratio linearly increases with the value of [H]/[H$_2$] (see also fig. 2.3.4). Thus, the upper limit of the method depends on the ability to measure the intensity of the molecular line, i.e. in general on the sensitivity of the detection system.

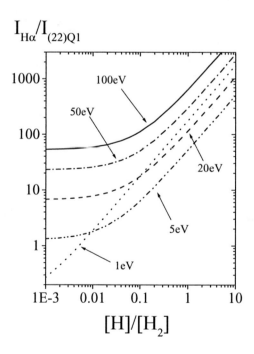

fig 2.3.5. The ratio of intensities of H$_\alpha$ and (2-2)Q1 lines I$_{H\alpha}$/I$_{(22)Q1}$ as function of the density ratio of atomic and molecular hydrogen [H]/[H$_2$], calculated for *case 2* at various T_e^{eff} and T$_g$ = 1000 K.

The ratio of intensities of H_α and (2-2)Q1 as function of T_e^{eff} calculated for various ratios of concentrations of atomic and molecular hydrogen (k=[H]/[H_2]) at a gas temperature 1000 K is shown in fig. 2.3.6. Here the intensity ratio has a minimum as function of T_e^{eff}. Therefore, the figure can be used for the determination of the maximum value of the density ratio. The determination of the parameter T_e^{eff} from measured intensity ratio $I_{H\alpha}/I_{(2-2)Q1}$ seems to be ambiguous.

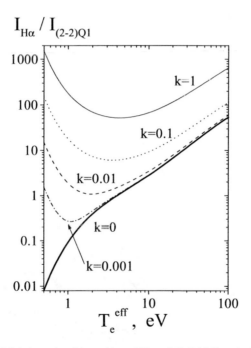

Fig. 2.3.6 the ratio of intensities of H_α and (2-2)Q1 lines $I_{H\alpha}/I_{(22)Q1}$ as function of T_e^{eff} calculated for various ratios of densities of atomic and molecular hydrogen (k = [H]/[H_2]) at a gas temperature of 1000 K.

2.3.3. Combination of two intensity ratios

Two independent measured ratios of intensities, $I_{H\alpha}/I_{H\beta}$ and $I_{H\alpha}/I_{(2-2)Q1}$, can be used for the determination of two unknown plasma parameters: (i) the required density ratio $[H]/[H_2]$ and (ii) the adjusted parameter T_e^{eff}, which describes in part the high energy tail of the EEDF (see eqs.2.3.1, 2.3.2). Using figs. 2.3.2, 2.3.3, 2.3.5 and 2.3.6 it is possible to determine at which numerical values of $[H]/[H_2]$ and T_e^{eff} the intensity ratios are sensitive to the plasma conditions. Those values are visualized in fig. 2.3.7. As one may see the plane $[H]/[H_2]$ and T_e^{eff} is divided in four regions.

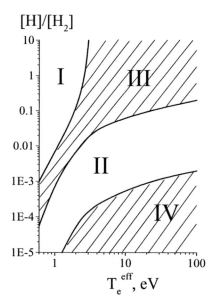

Fig. 2.3.7 Range of applicability of intensity ratios of H_α, H_β, and (2-2)Q1 lines for determination of the degree of dissociation of hydrogen. The areas I and II can be used; the areas III and IV – not.

Regions I and II can be used for the determinations of the degree of dissociation from intensity ratios.

In region I :

(i) both equations (2.3.1, 2.3.2) are sensitive to variations of the parameter T_e^{eff} (see fig. 2.3.3, 2.3.6),

(ii) equation (2.3.2) is sensitive to variations of the ratio [H]/[H$_2$] (fig. 2.3.5).

In region II :

(i) both equations (2.3.1, 2.3.2) are sensitive to variations of the parameter T_e^{eff} (see fig. 2.3.3, 2.3.6),

(ii) equation (2.3.1) is sensitive to variations of the ratio [H]/[H$_2$] (fig. 2.3.2).

Regions III and IV cannot be used for determinations of the degree of dissociation from intensity ratios under consideration.

In region III :

Eq. (2.3.1) does not depend on changes of the ratio [H]/[H$_2$] either of the parameter T_e^{eff} .

The eq. (2.3.2) is sensitive to a variation of [H]/[H$_2$] and with the smallest sensitivity to variation of T_e^{eff} due to a minimum of eq. (2.3.2) as function T_e^{eff} in this region. So the ratio of atomic and molecular lines, eq. (2.3.2), can be used for the determination of the degree of dissociation together with some other intensity ratios, which are depend on T_e^{eff} .

In region IV :

Both equations (2.3.1, 2.3.2) do not depend on changes of the ratio [H]/[H$_2$]

The regions I and II are comfortable for the determination of the degree of dissociation, and are separated by region III. In region III both intensity ratios, eq. (2.3.1, 2.3.2), as functions of T_e^{eff} have minima. So, the region I corresponds to the situation when both intensity ratios decrease with increasing T_e^{eff} at constant [H]/[H$_2$], while region II corresponds to the situation when both intensity ratios increase with T_e^{eff} , see figs. 2.3.3 and 2.3.6. Therefore, at a fixed value of the intensity ratios and [H]/[H$_2$] two values of T_e^{eff} can be found at special experimental conditions.

In the range of plasma parameters of region I the interpretation of the intensity ratios strongly depends on the knowledge of the behavior of the cross sections at the threshold region. At present this knowledge is not sufficient [15], and new theoretical data, special for Balmer lines, and experimental data, special for molecular hydrogen lines, about their threshold behavior of the cross section are necessary for further development of the emission spectroscopic technique.

In region II for the determination of the degree of dissociation special experimental conditions are required: a plasma with extremely low degree of dissociation or with a high density of fast electrons.

Thus, in the framework of the simple excitation deactivation model, used in this study, the set of equations (2.3.1, 2.3.2) can be applied to determine of the degree of dissociation of hydrogen in a wide range of plasma conditions. It is necessary to emphasize that:

(i) the determination could be ambiguous in some ranges of plasma conditions (see fig.2.3.3, 2.3.6, 2.3.7), and

(ii) the method cannot be applied in plasmas with a high degree of dissociation, where the tail of the EEDF is described by high values of the parameter T_e^{eff} (see figs. 2.3.2, 2.3.3, 2.3.7). In such situations the ratio of atomic line intensities is around 5 ($I_{H\alpha}/I_{H\beta} \cong 5$).

3. EXPERIMENTAL RESULTS AND DISCUSSIONS

3.1 Plasma sources and optical systems

The applicability of the method of determination of the degree of dissociation of hydrogen has been approved in three types of non-equilibrium discharges containing hydrogen:

a) a capillary DC-arc,

b) a RF, $f = 200$ kHz,

c) a MW, $f = 2.45$ GHz.

For the DC case, the plasma of a standard lamp, DVS-25, for investigation of hydrogen molecule spectra was used. It is a hot cathode capillary arc discharge (2 mm diameter) surrounded by metallic surfaces. The pressure of the gas filling, pure hydrogen, was 8 mbar and the current in the range between 50 and 300 mA. The set up is shown in fig. 3.1.1

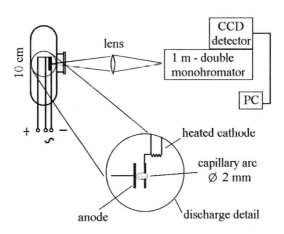

Fig.3.1.1 Experimental set-up for the DC arc discharge, DVS-25.

A 1 m Czerny-Turner double monochromator with a grating of 1800 grooves per mm in first order was used (Jobin Yvon, U1000). The CCD detector, pixel size of 19x19 μm^2 with 1024x512 pixels, Princeton Applied Research, was controlled by a personal computer, PC.

The RF discharge (200 kHz) was maintained in a 2-meter glass tube, diameter: 3.5cm, with one inner metallic surface, a grounded cylindrical Ni electrode located at one end of the tube, see fig. 3.1.2. All other cylindrical electrodes, two grounded and two for applying RF power, were located at the outer surface of the glass tube at a distance of 30 cm from each other. The radiation of the discharge, coming out perpendicular to the tube axis, was projected on to the entrance of a light cable by a glass lens with a focal length of 8 mm. The diameter of the lens was not exceeding 2 cm adjusted by a diaphragm. The lens was situated at a distance of 33 cm from the tube, so a size 2 mm of the projected image, which was cut by entrance diaphragm of the light cable in accordance to a size 0.5 cm of the light source. The second end of the light cable was connected to the entrance slit of a spectrometer. A 0.5 m Czerny-Turner spectrometer with a grating of 2400 grooves per mm in the first order was used (Acton Research Corporation, SpectraPro-500). A CCD detector, pixel size of 19x19 μm^2 with 512x512 pixels, Princeton Applied Research, was located in the focal plane of the spectrometer.

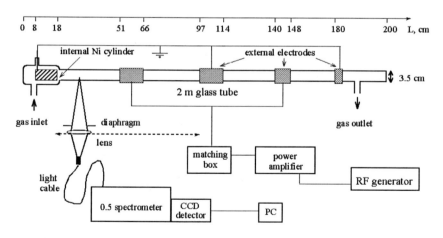

Fig. 3.1.2 Experimental setup for the RF discharge.

The signal of a RF generator (GF 21, RC – generator, Präcitronic) was connected to a power amplifier, model 1040L, ENI. A matching box, model TR1140-5G, ENI, provided the signal to external cylindrical electrodes of the discharge tube. The matching box contains a variable inductance. Capacitive unbalances were compensated due to the change of size and the position of the external electrodes.

The discharge was excited in pure hydrogen at a pressure 0.3-1 mbar and RF power of 170W.

It is known that at low pressures the volume association of atoms to molecules is negligible in comparison to that taking place at the surfaces. Metals are much more efficient than glass [59], [60]. Therefore the geometry of the RF discharge tube has been expected to provide a high gradient of the distribution of the degree of dissociation, from almost zero inside the inner electrode up to a certain level far from the electrode, under almost the same conditions along the plasma column.

The third plasma under study was exited in a planar microwave reactor [61] (see fig 3.1.3). The internal dimensions of the reactor vessel: length, height and width were 50, 25 and 35 cm, respectively. The microwave power (f= 2.45 GHz) was coupled into the plasma from the top by a special waveguide system through a quartz window transparent for microwaves. The plasma was mainly excited in the top part of the vessel just below the microwave window. The experimental set-up was the same as in [62, 63]. Gas flows of Ar, H_2 and of a B_2H_6-H_2 mixture were measured by separate mass flow controllers, and then the gases were mixed before entering the reactor. They were pumped out via a port in the reactor wall. The measurements were performed in a gas mixture of B_2H_6: H_2: Ar=3:63:33 with a total flow of 150sccm at a pressure of 2.5 mbar and microwave power of 1.5-3.5 kW.

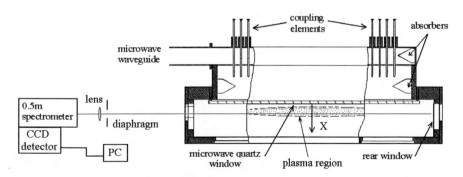

Fig.3.1.3 Experimental setup for MW discharge.

The plane of the rear optical window of the plasma reactor was imaged on the plane of the entrance slit of the spectrometer by an achromatic quartz lens with a focal length of $f = 300$ mm. The distance between the lens and the rear window was 140 cm. The length of plasma column was 49 cm. The solid angle $\Delta\Omega$ for the light from plasma being able to come into the spectrometer and to be recorded by the detector was adjusted by a diaphragm. The measurements have been performed with diaphragm diameter 5 mm corresponding to a negligibly small solid angle (5/140). Thus, the spatial resolution of the emission measurements within plasma column was better than 1.8 mm (5x49/140). The spectrometer and the CCD detector were the same as used for the RF-discharge.

Examples of the recorded spectra and identification are given in the appendix.

3.2 Determination of the spectral sensitivity

For the relative intensity calibration of the emission spectroscopic system, the determination of the spectral distribution of the sensitivity, measurements of a specially calibrated tungsten ribbon lamp were used. The signal generated by one column of pixels of the detector by the tungsten lamp during exposure time τ_{BL}, s, is:

$$K_{BL}(\lambda) = \sigma(\lambda)\rho(\lambda,T)B_0(\lambda,T)\frac{d\lambda}{dx}l\,S_{BL}\Delta\Omega_{BL}\tau_{BL} \qquad 3.2.1$$

where $K_{BL}(\lambda)$ is the signal given by the detector expressed in counts per pixel column, $\sigma(\lambda)$ is the quantum efficiency of the detector, counts per photon, $B_0(\lambda,T)$ is the spectral distribution of the black body radiation calculated in $\frac{photon}{nm\cdot str\cdot s\cdot cm^2}$, $\rho(\lambda,T)$ is the emissivity coefficient of tungsten, $d\lambda/dx$ is the linear dispersion of the spectrometer, nm per pixel column, l is the linear dimension of a pixel, cm, S_{BL} is the area of the band detected by the system, cm^2, and $\Delta\Omega_{BL}$ the solid angle, str, T is a temperature of the tungsten band, K.

For calibration of the presented systems a relative measurements were carry out:

$$\frac{K_{BL}(\lambda)}{K_{BL}(\lambda_{656})} = \frac{\sigma(\lambda)\rho(\lambda,T)B_0(\lambda,T)\dfrac{d\lambda}{dx}}{\sigma(\lambda_{656})\rho(\lambda_{656},T)B_0(\lambda_{656},T)\left(\dfrac{d\lambda}{dx}\right)_{656}} \qquad 3.2.2$$

The signal of the detector as function of the wavelength has been normalized on a signal at a wavelength of 656 nm.

In the case of a plasma column having the length L_{PL}, cm, and cross section S_{PL}, cm^2, the signal generated by one column of pixels may be presented as follows:

$$K_{PL}(\lambda) = \sigma(\lambda)\varepsilon(\lambda)\frac{d\lambda}{dx}lS_{PL}L_{PL}\frac{\Delta\Omega_{PL}}{4\pi}\tau_{PL} \qquad 3.2.3$$

where $K_{PL}(\lambda)$ is the signal given by the detector, $\varepsilon(\lambda)$ is plasma emissivity, $\frac{photon}{nm\cdot cm^3\cdot s}$, τ_{PL} is the exposure time to the plasma.

From eqs. 3.2.2 and 3.2.3 the plasma emissivity can be express as:

$$\varepsilon(\lambda) = \frac{1}{G_{PL}\,\tau_{PL}}\frac{K_{PL}(\lambda)}{S(\lambda)} \qquad 3.2.4$$

where $G_{PL} = lS_{PL}L_{PL}\dfrac{\Delta\Omega_{PL}}{4\pi}$ is a geometrical factor, $S(\lambda) = \dfrac{K_{BL}(\lambda)}{K_{BL}(\lambda_{656})}\dfrac{\rho(\lambda_{656},T)B_0(\lambda_{656},T)}{\rho(\lambda,T)B_0(\lambda,T)}$ is the relative sensitivity.

To find the relative sensitivity the function $K_{BL}(\lambda)/K_{BL}(\lambda_{656})$ has been measured for three currents of the band lamp, which correspond to calibrated brightness temperatures of the band lamp. The difference of the relative sensitivity values obtained for three different brightness temperatures was found to be less than 1% in wide spectral range. Experimental errors increase at wavelengths shorter than 300nm, because the intensity of the band lamp is decreasing sharply with the decreasing of the wavelengths. The relation between the brightness temperature and the black body temperature T can be found in [64]. Emission coefficients of tungsten are discussed in [65]. For the present work calculated emissivity values of tungsten band lamp have been used [66].

It is necessary to note that the relative sensitivity strongly depends on the alignment of the monochromator. The relative sensitivity of the 0.5-meter spectrometer, 2400 g/mm with two types of adjustment of the last mirror of the monochromator, which focus spectra on CCD matrix, is shown in fig. 3.2.6. One may see that the relative sensitivity strongly depends on the wavelength. In 2 nm wavelength range the sensitivity can be interpolated as a straight line. Readjustment of the monochromator mirror can change the relative sensitivity by a factor of more than two.

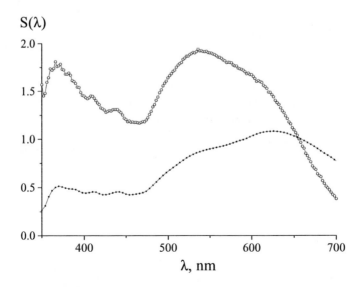

Fig. 3.2.6 The relative sensitivity of the 0.5 meter spectrometer with two types of alignment of the last mirror of the monochromator, which focuses light the on CCD matrix.

3.3   Method of the gas temperature determination

To determine the gas temperature the approach first proposed in [67] and later developed in [23, 28, 62, 63, 68, 69] has been used in the present study. The method is based on measured rotational distribution of intensity of molecular bands. In the present work only a short description of the method and technical routines are given for the Fulcher-α (2-2)Q band system of hydrogen.

The following assumptions have been used:

i). The population of rotational sublevels of the electronic and vibrational ground state, $X\Sigma_g^+$, v=0, N, obeys a Boltzmann distribution with temperature T_{rot}^{X0}, which equal to the gas temperature T_g.

ii). The excited states, $d^3\Pi_u^-$, v', N', are populated by electron impact from $X\Sigma_g^+$, $v=0$, N state and are depopulated by spontaneous decay.

iii). No change in the angular momentum of the molecule, $\Delta N=0$, occurs in electron impact excitation, $N=N'$.

Taking into account assumption i), the population density of a certain sublevel n_{X0N} of the ground state depends on the rotational energy E_{X0N} of the ground state according to:

$$n_{X0N} = n_{X00} g_{a.s}(2N+1)\exp(-\frac{E_{X0N}}{T_g}) \ , \qquad\qquad 3.3.1$$

where $g_{a.s}$ is the statistical weight of the X, 0, N sublevel of the homonuclear molecule depending on the symmetry in respect to permutation of the nuclear spin (for $d^3\Pi_u^-$ and $X\Sigma_g^+$ states of H_2 is valid: even rotational sublevels have $g_{a.s} = g_a = 1$, odd rotational sublevels have $g_{a.s} = g_s = 3$). The rotational energy and gas temperature are expressed in K.

Taking into account assumptions ii) and iii) the balance equation for excitation and deactivation of the $d^3\Pi_u^-$, v', N molecular sublevel can be written as:

$$\frac{n_{dv'N}}{\tau_{dv'N}} = n_e\, n_{X0N}\, K_{X0N}^{dv'N} \ , \qquad\qquad 3.3.2$$

where $n_{dv'N}$ and $\tau_{dv'N}$ are population density and radiative lifetime of the $d^3\Pi_u^-$, v', N sublevel; n_e is density of electrons; $K_{X0N}^{dv'N}$ is the rate coefficient for electronic excitation.

In work [23] it is discussed that $\tau_{dv'N}$ and $K_{X0N}^{dv'N}$ have a very weak dependence on N. So, substitute eq. 3.3.1 in eq. 3.3.2 and combine all terms, which do not depend on N, in *const* :

$$n_{dv'N} = const\ (2N+1)\exp\left(-\frac{E_{X0N}}{T_g}\right).$$ 3.3.3

The *const* is depending on temperature because the term n_{X00} is included, but it does not influence the relative population density and the determination of the gas temperature.

The diagonal Q branch of the Fulcher-α system occurs due to the radiative transitions $d^3\Pi_u^-$, v', $N \rightarrow a^3\Sigma_g^+$, v', N. The expression for the intensity of those lines can be written as:

$$I_{av'N}^{dv'N} = n_{dv'N} A_{av'N}^{dv'N}\ .$$ 3.3.4

Neglecting the vibrational-rotational interaction, the radiative transition probability of spontaneous emission $A_{av'N}^{dv'N}$ can be given as:

$$A_{av'N}^{dv'N} = \frac{64\pi^4}{3h}\left(v_{av'N}^{dv'N}\right)^3\ \left|R_e^{da}\right|^2 \frac{q_{v'v'}\,H_{NN}}{(2N+1)},$$ 3.3.5

where $/R_e^{da}/^2$ is the square of the matrix element of the dipole moment; $q_{v'v'}$ is the Franck – Condon factor; H_{NN} is the Hönle-London factor; $v_{av'N}^{dv'N}$ is the wavenumber of the line; and π and h are known constants.

For simple cases of the angular momentum coupling Hönle-London factors are known as analytic functions of rotational quantum numbers, depending only on the type of the electronic transition. In particular, for the $\Pi \rightarrow \Sigma$ electronic transition of the Q branch:

$$H_{NN} = 2N+1$$ 3.3.6

Substituting eq. 3.3.3 and eq. 3.3.5 in eq. 3.3.4, taking into account eq. 3.3.6, combine all factors, which do not depend on the rotational quantum number, in *const*, the reduced intensity of rovibronic line could be express as:

$$\tilde{I}_{av'N}^{dv'N} = \frac{I_{av'N}^{dv'N}}{g_{a.s}(2N+1)\left(v_{av'N}^{dv'N}\right)^3} = const\ \exp\left(-\frac{E_{X0N}}{T_g}\right)$$ 3.3.7

or

$$\ln\left(\tilde{I}_{av'N}^{dv'N}\right) = -\frac{1}{T_g}E_{X0N} + const$$ 3.3.8

Thus, the determination of the gas temperature requires to plot the dependence of the logarithm of the reduced line intensities on the molecular rotational energy in the ground state. All necessary data for the determination of the gas temperature from the relative intensities of the Fulcher-α (2-2)Q band system are given in table 3.1.

It should be emphasized, that the basic mechanism for establishing a Boltzmann distribution over the rotational levels in the ground state is the energy exchange between the translational and rotational degrees of freedom occurring in molecule-molecule collisions. For the coincidence of the temperatures corresponding to Maxwell and Boltzmann distributions, it is necessary, that

$$E_{X0N} - E_{X0N-2} < T_g \qquad\qquad 3.3.9$$

in the case of hydrogen molecule, when the transitions between neighbouring rotational levels are forbidden. That is, because of the exponential drop in the concentration of molecules with translational energy sufficient for the excitation of the N level from the $(N-2)$ level of molecule.

Table 3.1   Hydrogen molecular constant, which are necessary for the determination of the gas temperature from the relative intensities of the Fulcher-α (2-2)Q band system.

N	$2N+1$	$g_{a.s}$	$\left(\nu_{a2N}^{d2N}\right)^3$, 10^{-12} cm^{-3}	E_{X0N}, K	$E_{X0N} - E_{X0N-2}$, K
1	3	3	4.14248	170.5	−
2	5	1	4.13163	509.8	−
3	7	3	4.11549	1015.1	850
4	9	1	4.09427	1681.6	1180
5	11	3	4.06815	2503.8	1500
6	13	1	4.03738	3474.3	1800
7	15	3	4.00237	4586.2	2100
8	17	1	3.96332	5829.5	2400
9	19	3	3.92062	7196.7	2600
10	21	1	3.87466	8676.9	2900
11	23	3	3.82608	10261.3	3100
12	25	1	3.77467	11939.8	3300
13	27	3	3.72095	13702.6	3500

It is not easy to fulfill the inequality 3.3.9 in the case of the H_2 molecule. In table 3.1 the values of $E_{X0N} - E_{X0N-2}$ are given. As one may see, the gas temperature can be derived from the ratio of the rovibronic lines Q1 and Q3 when the gas temperature higher than 850 K. To involve Q2 and Q4 lines the gas temperature should have a value of 1200 K.

In fig. 3.3.1 the logarithm of the reduced intensity distributions of the Fulcher-α (2-2)Q band system, measured in the DVS-25 lamp at discharge currents of 300 mA and 50 mA are shown. The size of the points reflects the random error of the measured intensities. The first four points of the distribution were fitted by linear functions. From the slopes of the lines the temperatures have been derived, see eq. 3.3.8. Error bars of obtained values of the gas temperatures are a standard deviation of the fitting. In the distribution, which corresponds to 1350 K, first six points can be described by a Boltzmann distribution with the same temperature. The differences $E_{X06} - E_{X04} = 1800$ K and $E_{X05} - E_{X03} = 1500$ are higher then the obtained gas temperature. For the second distribution the situation is similar, $E_{X04} - E_{X02} = 1200$ K, at a determined temperature of 940 K.

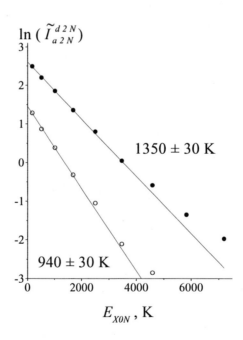

Fig. 3.3.1 Logarithm of the reduced intensity distributions of Fulcher-α (2-2)Q band system, see eqs.3.3.7 and 3.3.8, measured in DVS lamp at two different discharge currents.

3.4. Measured relative intensities of the spectral lines and gas temperature

Complete sets of experimental data, intensity ratios and gas temperature, which are necessary to solve the equation system (eqs.2.3.1 and 2.3.2) and for the evaluation of the degree of dissociation of hydrogen, measured in the present study, are shown in figs. 3.4.1-3.4.3: fig. 3.4.1 for the DC, fig. 3.4.2 for the RF and fig. 3.4.3 for the MW discharges.

The intensity ratios of Balmer lines H_α and H_β measured in three types of discharges are shown in figs. 3.4.1a, 3.4.2a and 3.4.3a. The intensity ratios are higher than 7 for all experimental conditions. Therefore, the ratio of atomic lines is in the range of the applicability of the method (see discussion in chapter 2.3.1).

In the case of the DC discharge (fig. 3.4.1a) the increase of the discharge current of a factor of 6 leads to a relatively small increase of the intensity ratio of about 10%. As one may see from the fig. 2.3.3 the increase of the intensity ratio with discharge current can be caused only by changes of the shape the EEDF. In contrast, the increase of the atomic density should lead to a decrease of the intensity ratio (fig. 2.3.2).

The intensity ratio of atomic lines clearly depends on the pressure in the case of the RF discharge (fig. 3.4.2 a). A pressure variation at about 3 times, from 0.3mbar up to 1mbar, leads to 30% changes of the intensity ratio. The dependence of the intensity ratio on the distance from the inner metallic electrode shows for all pressures the same behavior.

In the microwave discharge (fig. 3.4.3a) the intensity ratio of the Balmer lines shows opposite tendencies with increasing of microwave power for two distances of the optical axis from the microwave window.

In the present experiments the intensity ratio of the Balmer lines H_α and H_β was found to be in the range between 7.7 and 18, a change of more than factor 2. Therefore, this intensity ratio is sensitive to variations of the discharge type and of experimental conditions.

The intensity ratio of the Balmer line H_α and the molecular (2-2)Q1 line of the Fulcher-α band system measured in the three types of discharges is shown in fig. 3.4.1b, 3.4.2b and 3.4.3b. The intensity ratios of atomic and molecular lines strongly depend on the discharge conditions in the plasma sources under study. In the DC discharge (fig. 3.4.1b) the intensity ratio increase linearly with current by a factor of 2.5. The strongest variation of the intensity ratio was found in the RF discharge (fig. 3.4.2 b). For a pressure of 0.3 mbar the intensity ratio grows with the distance from the inner metallic electrode by two orders of magnitude: form 3.7 up to 360. From fig. 3.4.2 b it can be seen that the intensity ratio also

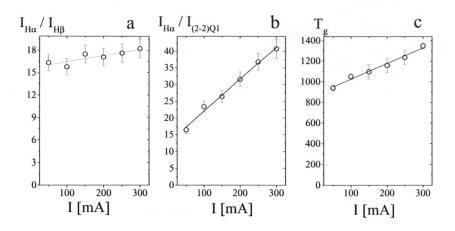

Fig. 3.4.1 The intensity ratios of H_α and H_β lines (a), H_α and (2-2)Q1 lines (b), and the gas temperature are measured in the DC arc discharge versus the discharge current

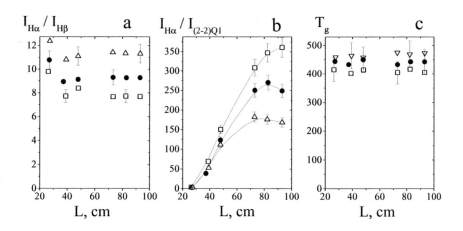

Fig. 3.4.2 The intensity ratios of H_α and H_β lines (a), H_α and (2-2)Q1 lines (b), and the gas temperature are measured in the RF discharge versus the distance from inner metallic electrode for different pressures: ($\square$) – 0.3 mbar; ($\bullet$) – 0.6 mbar; ($\triangle$) – 1 mbar.

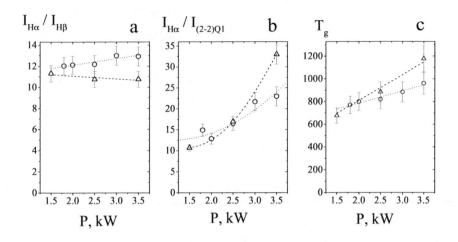

Fig.3.4.3 The intensity ratios of H_α and H_β lines (a), H_α and (2-2)Q1 lines (b), and the gas temperature are measured in the microwave discharge versus mw power at two distances of the optical axis from mw window: ($\bigcirc$) – 4cm; ($\triangle$) – 2cm.

clearly depends on the gas pressure. In the MW discharge (fig. 3.4.3b) the intensity ratio related to the small distance of 2 cm from the MW window increases more pronounced with MW power than for the bigger distance of 4 cm.

As it was mention above, the intensity ratio of atomic and molecular lines strongly depends on the gas temperature, see eq. 2.3.2 and fig. 2.1.2. Obtained results of the temperature determination are shown in figs. 3.4.1c, 3.4.2 c and 3.4.3c. One may see that different types of the discharges are characterized by in different temperature ranges: the DC arc discharge, fig. 3.4.1c, at 900 to 1300K, the RF discharge, fig. 3.4.2 c, at 400 to 500K and the MW discharge (fig. 3.4.3c) at 700 to 1200K. Moreover the gas temperature depends on the discharge conditions and influences qualitatively the relation between the intensity ratio and the degree of dissociation. An increase of the gas temperature leads to an increase of the intensity ratio, see eq.2.3.2 and fig.2.1.2. Therefore, account of the gas temperature in interpretation of the molecular line intensity is important for quantitative and qualitative results of determination of the degree of dissociation.

3.5. Determination of the degree of dissociation of hydrogen

3.5.1 Description of the method

In order to get the degree of dissociation of hydrogen from measured intensity ratios and gas temperature one needs to solve equation system eqs. (2.3.1) and (2.3.2). For this purpose it is suggested to express the ratio of atomic and molecular densities as function of the parameter T_e^{eff} directly from eqs.(2.3.1) and (2.3.2):

$$\frac{[H]}{[H_2]} = \frac{K_{dis}^{em}(T_e^{eff}\,|\,H_\alpha) - \dfrac{I_{H\alpha}}{I_{H\beta}} K_{dis}^{em}(T_e^{eff}\,|\,H_\beta)}{\dfrac{I_{H\alpha}}{I_{H\beta}} K_{dir}^{em}(T_e^{eff}\,|\,H_\beta) - K_{dir}^{em}(T_e^{eff}\,|\,H_\alpha)} \tag{3.5.1}$$

$$\frac{[H]}{[H_2]} = \frac{I_{H\alpha}}{I_{(2-2)Q1}} \frac{A_{a21}^{d31}\tau_{d21}K_{mol}^{ex}(T_e^{eff}\,|\,d^-21 \leftarrow X01)}{K_{dir}^{em}\left(T_e^{eff}\,|\,H_\alpha\right)}\eta(T_g) - \frac{K_{dis}^{em}(T_e^{eff}\,|\,H_\alpha)}{K_{dir}^{em}(T_e^{eff}\,|\,H_a)}, \tag{3.5.2}$$

or from the ratio of eqs.(2.3.1) and (2.3.2):

$$\frac{[H]}{[H_2]} = \frac{I_{H\beta}}{I_{(2-2)Q1}} \frac{A_{a21}^{d31}\tau_{d21}K_{mol}^{ex}(T_e^{eff}\,|\,d^-21 \leftarrow X01)}{K_{dir}^{em}\left(T_e^{eff}\,|\,H_\beta\right)}\eta(T_g) - \frac{K_{dis}^{em}(T_e^{eff}\,|\,H_\beta)}{K_{dir}^{em}(T_e^{eff}\,|\,H_\beta)} \tag{3.5.3}$$

If measured quantities are inserted in eqs.(3.5.1-3.5.3) than every equation represent a curve on the plain T_e^{eff}, [H]/[H_2] and a point of intersection of the curves is the solution of the system. The examples of those curves are calculated from the intensity ratios and gas temperature, which were measured in three different discharges, are shown in fig. 3.5.1. The rate coefficients of the excitation of Balmer lines corresponding to *case 2*, i.e. Boltzmann distribution of the population density over fine structure sublevels, were used.

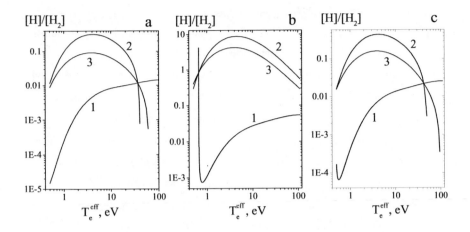

Fig. 3.5.1 Examples of finding the density ratio of atomic and molecular hydrogen from the measured intensity ratios of spectral lines for a) DC (i = 50 mA), b) RF (p = 0.6 mbar, L = 58.1 cm), c) MW (at distance of x = 4 cm, P = 3 kW, p = 2.5 mbar) discharges. The curve 1 corresponds to equation (3.5.1), curve 2 – to eq.(3.5.2), 3 – (3.5.3).

One may see from the fig. 3.5.1 that for the present examples the curves can intersect in two regions: for low values of the parameter T_e^{eff} (< 1 eV) or for high values of parameter T_e^{eff} (> 20 eV). At the region of low T_e^{eff}, eq. (3.5.1) is too sensitive to variations of the parameter T_e^{eff} (see fig 3.5.1b, curve 1) and it is difficult to calculate the curve in this region. To find the solution of the system for low values of the parameter T_e^{eff} the eqs. (3.5.2) and (3.5.3) have been used, shown in curves 2 and 3. At the region of high T_e^{eff} the curves 1 and 2, corresponding to eqs. (3.5.1) and (3.5.2), have a angle of intersection near to 90° and therefore have been chosen for determination of the degree of dissociation.

The parameter T_e^{eff} was varied from 0.5 eV to 100 eV. The lower limit of the parameter T_e^{eff} has been chosen for two reasons: (i) for low values of parameter T_e^{eff} the rate coefficients are mainly determined by the threshold behavior of the cross sections, which is unknown and (ii) the rate coefficients are dramatically depend on the value of the T_e^{eff}, i.e. from the shape of the EEDF.

One may see from table 4 in the appendix, that between the values of T_e^{eff} 0.5 eV and 1 eV emission rate coefficients of the H_α line for direct electron impact excitation changes six orders of magnitude, while between 1 eV and 2 eV the cross section changes only two orders of magnitude. The other rate coefficients listed in table 4 and 5 have even stronger dependencies on T_e^{eff} in this region. So, for low values of T_e^{eff} intensity ratios are sensitive to mistakes in value of T_e^{eff}, which could be caused by non-Maxwellian behavior of the EEDF. Therefore such mistakes in the value of T_e^{eff} together with insufficient knowledge of the cross sections can lead to noticeable mistakes in the determination of the degree of dissociation, when the high energy tail of the EEDF is described by too low value of T_e^{eff}.

The upper limit of the parameter T_e^{eff} should also be chosen carefully, since an approximation of an EEDF above the excitation threshold of the spectral lines by a Maxwellian function is based on the assumption that the EEDF drops rapidly.

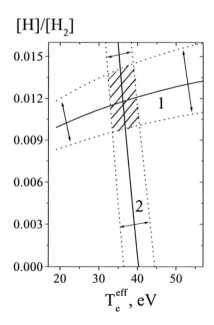

Fig. 3.5.2 Example of finding the density ratio of atomic and molecular hydrogen from the measured intensity ratios of spectral lines for the DC discharge, i = 50 mA, zoomed on the intersection region. The curve 1 corresponds to equation (3.5.1), curve 2 – to eq.(3.5.2).

An example of the intersection region of the curves corresponding to eqs. (3.5.1) and (3.5.2) for the DC arc discharge is shown in fig. 3.5.2. The dashed lines represent the errors due to the experimental uncertainties of measuring relative intensities and the gas temperature. In the picture is shown that the interception point has a certain area due to experimental errors. The size of this area has been used for the estimation of the errors of the measured density ratios and of the parameters T_e^{eff}. The results of the determination of the density ratio of atomic and molecular hydrogen and of the parameter T_e^{eff} are shown in figs. 3.5.3-3.5.7.

3.5.2 Consideration of the fine structure of Balmer lines

The application of the rate coefficients corresponding to *case 1*, when mixing of the population densities over the fine structure sublevels of hydrogen atoms is neglected, leads to negative results of the degree of dissociation for DC and MW discharges, i.e. it is without any physical meaning (not shown). This fact indirectly confirms that the *case 2*, when the fine structure sublevels of the hydrogen atoms are populated in accordance to the Boltzmann law, better corresponds to plasma experiments than *case 1*.

The solutions obtained in both limit *cases* for the RF discharge at a pressure of 0.6 mbar are shown in fig. 3.5.3. The dependence of the solutions on the distance from the inner metallic electrode is qualitatively the same, but the numerical values of the density ratio of atomic and molecular hydrogen differs by a factor of about 2.

So, results of the determination of the degree of dissociation obtained in the present work for three different reactors emphasize the importance of consideration of the H atom fine structure. The disregard of the redistribution of the population density over the fine structure sublevels leads to serious mistakes in determination of the degree of dissociation of hydrogen even to absurd results like negative values. In the figs. 3.5.4 – 3.5.7 results only for *case 2* for all three discharges are shown.

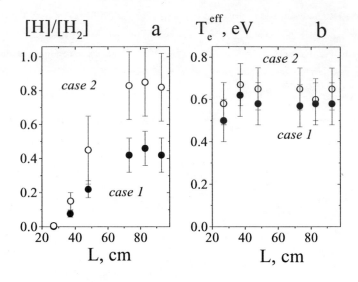

fig.3.5.3. The density ratio of atomic and molecular hydrogen (a) and effective parameter T_e^{eff} (b) versus distance from inner metallic electrode in RF discharge (0.6mbar) for two limit cases of excitation of Balmer lines. (●) - *case 1* (in accordance to beam experiments); (○) – *case 2* (in accordance to plasma experiments)

3.5.3 Results of the application of the method

3.5.3a   DC-arc discharge

The density ratio of atomic and molecular hydrogen together with the effective parameter T_e^{eff} obtained for conditions of the DC discharge are shown in fig. 3.5.4. It can be seen that the ratio [H]/[H₂] is about 0.012 and does not change with the discharge current. The lamp DVS-25, used here, is specially constructed for investigation of emission spectra of molecular hydrogen and has large metallic electrode surfaces. It can be expected that the density of hydrogen molecules is higher then the density of atoms. The rate of dissociation should increase with the discharge current, but could be compensated by an increasing flux of atoms to the metallic surfaces.

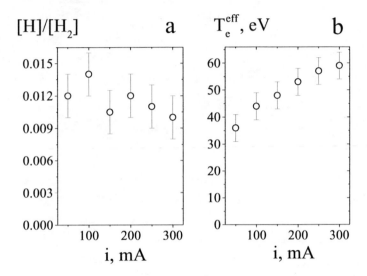

Fig. 3.5.4. The density ratio of atomic and molecular hydrogen (a) and effective parameter T_e^{eff} (b) versus current of the DC arc discharge.

The flux of the atoms to the surfaces could be estimated in framework of the elementary kinetic theory of gases, see for example [70]. The flux of particles, J, is proportional to product of the mean free path, λ, and the average velocity $\bar{v}$:

$$J \sim \lambda \cdot \bar{v} \frac{d[H]}{dx} ,$$ 3.5.4

where $d[H]/dx$ is a density gradient.

The average velocity of particles is proportional to the square root of the temperature:

$$\bar{v} \sim \sqrt{T_g} .$$ 3.5.5

The mean free path is proportional to the gas temperature if the pressure is constant. The plasma volume is much smaller then the volume of the lamp. So, it can be assumed that the pressure is constant:

$$\lambda \sim T_g .$$ 3.5.6

Assuming that (i) near to the metallic walls only hydrogen molecules exist and (ii) in the plasma the concentration of the atoms corresponds to the measured degree of dissociation, then the atomic gradient $d[H]/dx$ is constant for all experimental conditions due to the constancy of the measured degree of dissociation.

Therefore, the atomic flux to the walls depends only on the gas temperature in the present consideration. Substituting eqs. (3.5.5) and (3.5.6) into (3.5.4) for the atomic flux J_1, corresponding to the gas temperature T_1 and for J_2 corresponding to T_2, a ratio of fluxes can be obtained:

$$\frac{J_2}{J_1} = \frac{T_2}{T_1} \sqrt{\frac{T_2}{T_1}} .$$ 3.5.7

In the DVS-25 lamp the gas temperature grows up with discharge current from 900K till 1300K. Therefore the flux of atoms changes in 1300/900*sqrt(1300/900)=1.7 times.

The parameter T_e^{eff} shows rather high values (see fig.3.5.4b), what is not surprisingly, because it is known, that EEDFs of discharges with constrictions can have secondary maximums above the excitation thresholds.

3.5.3b   RF discharge

The density ratio of atomic and molecular hydrogen together with the effective parameter T_e^{eff} obtained for conditions of the RF discharge is shown in fig. 3.5.5. The ratio $[H]/[H_2]$ increases by two orders of magnitude with L while the parameter T_e^{eff} keeps constant for a certain pressure.

Here the parameter T_e^{eff} is rather small, about 0.5 - 0.8 eV. The rate coefficients calculated with such values of T_e^{eff} are mainly determined by the threshold behavior of the corresponding cross sections. The values of the rate coefficients, which were used in the present study and listed in tables 4 and 5 in the appendix, are based on the simplest assumption: the linear growth of the cross sections as function of the electron energy, from zero at the threshold up to its maximum. Nevertheless, the present method leads to reasonable results, as supposed from constructed geometry: the sharp growth of the degree of dissociation with increasing distance L from the inner metallic electrode, see chapter 3.1.

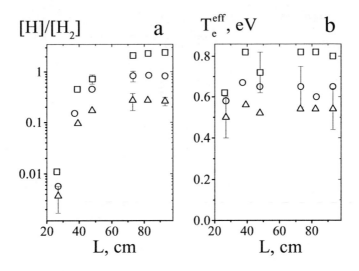

Fig. 3.5.5. The density ratio of atomic and molecular hydrogen and the effective parameter T_e^{eff} versus the distance from the inner metallic electrode in the RF discharge for different pressures: ($\square$) – 0.3 mbar; ($\bigcirc$) – 0.6 mbar; ($\triangle$) – 1 mbar.

3.5.3c   MW discharge.

For intensity ratios and the gas temperature measured in the MW discharge the equation system eqs. (2.3.1) and (2.3.2) gives two solutions (see fig 3.5.1c), which are shown in figs. 3.5.6 and 3.5.7. Both solutions show qualitatively the same results: (i) the density ratio increases with MW power at the distance of 2 cm of the optical axis from the MW window, while for a greater distance of 4 cm the variation of the density ratio does not exceed the experimental error; (ii) the parameter T_e^{eff} does not change with the distance of the optical axis from the MW window. The quantitative difference of the density ratio of two solutions is less than factor 2. The solution for low values of T_e^{eff} considerably depends on the behavior of the cross sections in threshold region, which is unknown. Therefore, the solution for higher values of T_e^{eff} seems to be more reliable.

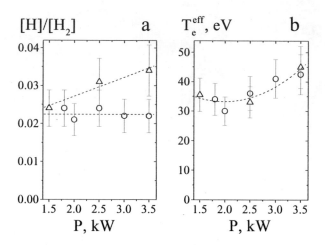

Fig. 3.5.6. The density ratio of atomic and molecular hydrogen and the effective parameter T_e^{eff} versus the microwave power at two distances of the optical axis from the MW window: ($\bigcirc$) – 4cm; ($\triangle$) – 2cm.
The Solution is corresponding to the high value of the parameter T_e^{eff} (see fig.3.5.1)

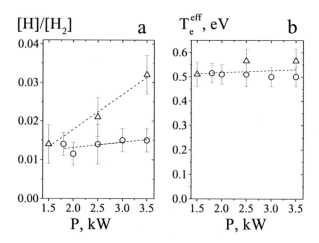

Fig. 3.5.7. The density ratio of atomic and molecular hydrogen and effective parameter T_e^{eff} versus microwave power at two distances of optical axis from mw window: ($\bigcirc$) – 4cm; ($\triangle$) – 2cm.
The solution is corresponding to the low value of the parameter T_e^{eff} (see fig.3.5.1)

4. SUMMARY AND CONCLUSIONS

The present work has been focused on the investigation of opportunities of a purely spectroscopic detection of the degree of dissociation of hydrogen in non-equilibrium plasmas. The general results of this study are related to two fields of physics:

    (I) Physics of elementary processes

    (II)   Plasma diagnostics

The principal original results in the field of physics of elementary processes can be formulated as follows:

(I.I)   The possibility of a correct consideration of the fine structure of the levels of the hydrogen atom in excitation and deactivation kinetics has been shown for two limit cases:

        Case 1, which can be observed in beam experiments, where the rates of the processes leading to a redistribution of the populations in the fine structure are negligible, and atoms emit from the sublevels into which they were excited,

        Case 2, which can be observed in plasma experiments, where the processes leading to a redistribution of the populations are dominant, and a Boltzmann distribution of populations over the *nlj*-sublevels is established.

(I.II)   The consideration of the fine structure of the levels of the hydrogen atom in excitation of the total intensities of the H_α and H_β lines has shown the following:

(i) The energy dependencies of the cross sections of the H_α and H_β line excitation by direct electron impact in the two limit cases differ significantly, and the ratios of the rate coefficients, calculated with a Maxwell electron energy distribution function, depend on the electron temperature.

(ii) Significant differences of the cross sections and rate coefficients have been found in the two limit cases, for direct and dissociative excitation of the H_α and H_β lines. For example, this difference reaches the factor of 15 for dissociative excitation of the H_β line.

(I.III)   The most reliable data on cross sections and rate coefficients of direct and dissociative excitation of the H_α and H_β lines by electron impact have been compiled in the two limit cases. In particular, it was shown that the rate coefficients provided by the ADAS European database [47] for levels with n=3 and 4 differ by a factor of 1.5 from the data of this study, which are based on the most accurate *ab initio* calculations of Bray [38] and which coincide well with the experimental data of the Gallagher team [30].

(I.IV) It was also shown that, among the four first, rather easily detectable, lines of the Balmer series, only the two first lines, H_α and H_β, can be currently used in plasma spectroscopy because of the absence of data on the partial cross sections of hydrogen atom nl-sublevels ($n \geq 5$) excitation.

This would require new data on the partial cross sections of direct excitation of levels with $n \geq 5$ by electron impact, as well as a more detailed experimental study of the partial cross sections of the dissociative excitation of levels with $n \geq 3$ as a function of the incident electron energy.

The principal original results in the field of plasma diagnostics can be formulated as follows:

(II.I) A new purely spectroscopic method of determination of the degree of dissociation of hydrogen has been suggested. The method is based on the measurement of relative intensities of spectral lines of atomic and molecular hydrogen and of the gas temperature. Principal differences of the suggested method from previous spectroscopic methods are:

(i) Consideration of the fine structure of the spectral lines of the Balmer series and of the ro-vibronic structure of the molecular bands.

(ii) The method is based on two or more pairs of intensity ratios of atomic and molecular lines.

(iii) The method does not need the measurement of the EEDF and the electron temperature.

(II.II) The range of the applicability of the method, related to use relative intensities of the H_α, H_β lines and of the (2-2)Q1 line of the Fulcher-α system, has been estimated.

(II.III) It has been experimentally proven that the relative intensities of the H_α, H_β and (2-2)Q1 spectral lines are sensitive to the variation of the experimental conditions.

(II.IV) The experimental approvement of the new method has been performed at three types of non-equilibrium discharges containing hydrogen:

    a) a capillary DC-arc discharge,

    b) a RF discharge, $f = 200$ kHz,

    c) a MW discharge, $f = 2.45$ GHz.

The results of the application of the method using relative intensities of the H_α, H_β and (2-2)Q1 spectral lines can be to concluded as follows:

(i)   The method clearly detects changes of the degree of dissociation, at least in the case of RF and MW discharges.

(ii)   The method gives reasonable absolute values of the degree of dissociation, at least for DC-arc and RF discharges, where near metallic walls domination of the molecular species is expected.

(iii)   The importance of considering the fine structure of the H atom for plasma diagnostics has been shown. The disregard of the redistribution of the population density over the fine structure sublevels leads to serious mistakes in determination of the degree of dissociation of hydrogen even to absurd results like negative values.

(II.V) It was shown that for further development of the suggested method it is necessary:

(i)   To get more detailed information about the threshold behaviour of cross sections. This information would help to extend the range of applicability of the method and to obtain more reliable results especially for plasmas where the high-energy tail of the EEDF can be approximated by a Maxwellian function with a temperature below 1 eV.

(ii)   To involve more than two ratios of intensities. Then, the method would have its own internal check, since various intensity ratios should give the same value of $[H]/[H_2]$.

These problems should be solved within physics of electronic and atomic collisions, which can provide new data on excitation cross sections of atomic and molecular hydrogen. This would open new possibilities of a reliable purely spectroscopic determination of the degree of dissociation of hydrogen in non-equilibrium plasmas.

ACKNOWLEDGEMENTS

First of all I would like to express my sincere gratitude to PD Dr. J. Röpcke and Prof. Dr. B.P.Lavrov (St.-Petersburg State University) for suggesting the topic of the present work and for their permanent attention and continuous support.

I would like to thank to M. Kalatchev, Dr. M. Käning, Dr. N. Lang and Dr. M. Osiac for useful discussions and for support of experimental part of the work.

Many thanks to D.Gött for technical assistance during experiments.

I am thankful to scientists and staff of INP-Greifswald for thier hospitality, friendliness and help in everyday life, especially to my colleagues with whom I worked in the group of PD Dr. J. Röpcke: S. Glitsch, D. Gött, Dr. Hempel, M. Hübner, Dr. N. Lang, Dr. L. Mechold, Dr. M. Osiac, S. Sass, G. D. Stancu and H. Zimmermann.

Present work was partly supported in the framework of the German-Russian-French Trilateral Cooperation Project of University of Greifswald, St.-Petersburg State University and University of Paris-South.

Hiermit erkläre ich, daß diese Arbeit bisher von mir weder an der Mathematisch-Naturwissenschaftlichen Fakultät der Ernst-Moritz-Arndt-Universität Greifswald noch einer anderen wissenschaftlichen Einrichtung zum Zwecke der Promotion eingereicht wurde.

Ferner erkläre ich, daß diese Arbeit selbständig verfaßt und keine anderen als die darin angegebenen Hilfsmittel benutzt habe.

6. REFERENCES

1. Lavrov B.P. and Simonov V.Y. 1987 *Determination of the degree of dissociation of hydrogen plasmas.* High Temp. **25** 482

2. Tankala K. and DebRoy T. 1992 *Modeling of the role of atomic hydrogen in heat transfer during hot filament assisted deposition of diamond.* J.Appl.Phys. **72** 712

3. Wagner D., Dikmen B. and Döbele H.F. 1998 *Comparison of atomic and molecular densities and temperatures of deuterium and hydrogen plasmas in a magnetic multipole source.* Plasma Sources Sci.Technol. **7** 462

4. Boogaarts M.G.H., Mazouffre S., Brinkman G.J., van der Heijden H.W.P., Vankan P., van der Mullen J.A.M., Schram D.C. and Döbele H.F. 2002 *Quantitative two-photon laser-induced fluorescence measurements of atomic hydrogen densities, temperatures, and velocities in an expanding thermal plasma.* Rev. Sci. Instrum. **73** 73

5. Redman S.A., Chung C., Rosser K.N. and Ashfold M.N.R. 1999 *Resonance enhanced multiphoton ionization probing of H atoms in a hot filament chemical vapour deposition reactor.* Phys.Chem.Chem.Phys. **1** 1415

6. Biel W., Bröse M., David M., Kempkens H. and Uhlenbusch J. 1997 *Determination of atomic and molecular particle densities and temperatures in low-pressure hydrogen hollow cathode discharge.* Plasma Phys. Control. Fusion **39** 661

7. Lavrov B.P., Melnikov A.S., Kaening M. and Röpcke J. 1999 *UV continuum emission and diagnostics of hydrogen-containing nonequilibrium plasmas.* Phys.Rev.E: **59** 3526

8. Abroyan M.A., Kagan Yu.M., Kolokolov N.B., Lavrov B.P. 1975 *On determination of dissociation degree of hydrogen in monoplasmatron.* ICPIG XII, Eindhoven, Nederlands, 101

9. Abroyan M.A., Demidov V.I., Kagan Yu.M., Kolokolov N.B., Lavrov B.P. 1975 *Spectroscopic study of the constricted arc discharge in hydrogen at low pressure.* Opt.Spectrosc. **39** 12

10. Devyatov A.M., Kalinin A.V. and Miiovich S.R. 1991 *Estimate of the degree of dissociation in a low-pressure hydrogen plasma from measured intensity ratios of the H_2 line of the Fulcher-a ($d^3\Pi_u$-$a^3\Sigma_g$) system and the H_a line.* Opt.Spectrosc. **71** 525.

11. Lang N., Kalatchev M., Kaening M., Lavrov B.P. and Roepcke J. 1999 *Time behaviour of various emissions in modulated hydrogen microwave discharge.* FLTPD III, Saillon, Switzerland, 253

12. Schulz-von der Gathen V. and Doebele H.F. 1996 *Critical comparison of emission spectroscopic determination of dissociation in hydrogen RF discharges.* Plasma Chem. Plasma Process. **16** 461

13. Lavrov B.P., Pipa A.V. and Röpcke J. 2003 *On spectroscopic determination of hydrogen dissociation degree in non-equilibrium plasmas of DC, RF and microwave discharges.* FLTPD V, Villaggio Cardigliano, Italy, 196

14.   Coburn J.W. and Chen M. 1981 *Dependence of F atom density on pressure and flow rate in CF4 glow discharges as determined by emission spectroscopy*. J.Vac.Sci.Technol. **18** 353

15.   Lavrov B.P. and Pipa A.V. 2002 *Account of the fine structure of hydrogen atom levels in effective emission cross sections of Balmer lines excited by electron impact in gases and plasma*. Opt.Spectrosc. **92** 647

16.   Lavrov B.P. and Prosikhin V.P. 1988 *Electronic excitation in hydrogen gas-discharge low-pressure plasma* Opt.Spectrosc. **64** 298

17.   Lavrov B.P., Solov'ev A.A. and Tyutchev M.V. 1980 *Populations of the rotational levels of the $d^3\Pi_u$, v states of H_2,HD and D_2 in an RF discharge* J. Appl.Spectrosc. **32** 316

18.   Condon E.U., Shortley G.H. 1959 *The theory of atomic spectra* (Cambridge, At the university press)

19.   Verolainen Ya.F. 1998 *Radiative lifetimes of excited states of I group atoms* Available from VINITI, No.1594-V98, St. Petersburg, (in Russian)

20.   Ortiz M. and Campos J 1980 *Emission of H_α and H_β lines excited by electron impact on H_2*. J.Chem.Phys **72** 5635

21.   Moore C.E. 1993 *Tables of Spectra of Hydrogen, Carbon, Nitrogen, and Oxygen Atoms and Ions*. ed Gallagher J.W. (CRC Press, Boca Raton)

22.   Vainshtein L.A., Sobel'man I.I. and Yukov E.A. 1973 *Sections of Atom and Ion Excitation by Electrons* (Moscow: Nauka)

23.   Lavrov B.P. 1980 *Determination of the gas temperatures of a low-pressure plasma from the intensity of the H_2 and D_2 molecular bands. III. Relationship between the intensity distribution in a band and the gas temperature*. Opt.Spectrosc., **48** 375

24. Landau L.D. and Lifshitz E.M. 1977 *Quantum mechanics: non-relativistic theory* (Pergamon press Ltd., third edition) (Moscow: Nauka, 1974)

25. Lavrov B.P., Ostrovskii V.N. and Ustimov V.I. 1979 *Rotation transitions in the excitation of electronic states of molecules by electron impact*. Sov.Phys.JETP **49** 772 (Zh.Eksp.Teor.Fiz. 1979 vol.76, p.1521)

26. Lavrov B.P., Ostrovskii V.N. and Ustimov V.I. 1979 *Determination of the gas temperatures of a low-pressure plasma from the intensity of the H_2 and D_2 molecular bands. II. Rotational transitions under electron impact excitation*. Opt.Spectrosc. **47** 30

27.   Lavrov B.P., Ostrovskii V.N. and Ustimov V.I. 1979 *Cross section and rate constants for electron-impact excitation of electronic-rotational molecular states*. Sov.Tech.Phys.Lett. **5** 142 (Pis'ma Zh.Tech.Fiz. **5** 355).

28.   Astashkevich S.A, Kaning M., Kaning E., Kokina N.V., Lavrov B.P., Ohl A. and Roepcke J. 1996 *Radiative characteristics of $3p\Sigma$, Π; $3d\Pi$, Δ^- states of H_2 and determination of gas temperature of low pressure hydrogen containing plasmas* J. Quant. Spectrosc. Radiat. Transfer **56** 725

29.   Landau L.D. and Lifshitz E.M. 1995 *Theoretical physic V: Statistical physics, part 1* (Moscow: Nauka)

30.   Mahan A.H., Gallagher A. and Smith S.J. 1976 *Electron impact excitation of the 3S, 3P, and 3D states of H2* Phys.Rev.A **13** 156

31.   Kleinpoppen H. and Kraiss E. 1968 *Cross section and polarization of the H_α line by electron impact excitation* Phys.Rev.Letters **20** 361

32.   Walker J.D.Jr. and John R.M.St., 1974 *Design of a high density atomic hydrogen source and determination of Balmer cross sections.* J.Chem.Physics **61** 2394

33.   Dunseath K.M., Terao-Dunseath M., Dourneuf M.Le. and Launay J-M. 1999 *Theoretical study of electron-impact excitation of H(1s) into the n ≤ 4 states using a two-dimensional R-matrix propagator* J.Phys.B **32** 1739

34.   King G.C., Trajmar S. and McConkey J.W. 1989 *Electron impact on atomic hydrogen – recent results and New Directions* Comments At. Mol. Phys. **23** 229

35.   Callaway J. and McDowell M.R.C. 1983 *What we do and do not know about electron impact excitation of atomic hydrogen* Comments At. Mol. Phys. **13** 19

36.   Odgers B.R., Scott M.P. and Burke P.G. 1994 *Electron scattering from hydrogen atoms at intermediate energies: integrated cross section to n=3 levels" intermediate energy R-matrix (IERM) approach* J.Phys.B **27** 2577

37.   Mansky E.J. and Flannery M.R. 1990 *The multichannel eikonal theory of electron-hydrogen collisions: I. Excitation of H(1s)* J.Phys.B **23** 4549

38.   Bray I. 1999 CCC On Line Database http://atom.murdoch.edu.au/CCC-WWW/

39.   Vainshtein L.A. 1965 *Effective excitation cross section in Born approximation* Opt. Spektrosk. **18** 947 (in Russian)

40.   Moehlmann G.R., de Heer F.J. and Los J. 1977 *Emission cross sections of Balmer-α, β, γ radiation for electrons (0-2000eV) on H_2 and D_2* Chem. Phys. **25** 103

41.   Karolis C. and Harting E. 1978 *Electron impact dissociation cross sections in hydrogen and deuterium, leading to Balmer alpha and beta emission* J.Phys.B **11** 357

42.   Moehlmann G.R., Tsurubuchi S. and de Heer F.J. 1976 *Excitation cross section for 3s, 3p, and 3d sublevels of atomic hydrogen split from simple molecules by high-energy electron impact* Chem. Phys. **18** 145

43.   Tsurubuchi S., Moehlmann G.R. and de Heer F.J. 1977 *Excitation cross sections for the H_{3s} state for electron impact on hydrogen containing molecules* Chem. Phys. Lett. **48** 477

44.   Yonekura N., Furuya K., Nakashima K. and Ogawa T. 1997 *Absolute emission cross section of the Paschen-α line and production dynamics of the 4F state of the excited hydrogen atom in e-H_2 collisions* J. Chem. Phys. **107** 1147

45.   Park C. 1971 *Electron impact excitation rate coefficients for hydrogen, helium and alkali atoms* J.Quant. Spectrosc. Radiat. Transfer. **11** 7

46.   Anderson H., Balance C.P., Badnell N.R. and Summers H.P. 2000 *An R-matrix with pseudostates approach to the electron-impact excitation of H_I for diagnostic applications in fusion plasmas* J.Phys.B **33** 1255

47.   Summers H.P. 1999 ADAS User Manual Version 2.1 http://adas.phys.strath.ac.uk/ University of Strathclyde, Glasgow.

48.   Benringer K. and Fantz U. 2000 *The influence of opacity on hydrogen excited-state population and applications to low-temperature plasmas* New Jour. Phys. **2** 23.1

49.   Moehlmann G.R. and de Heer F.G. 1976 *Emission cross section of the H_2 $(3p^3\Pi_u \rightarrow 2s^3\Sigma_g^+)$ transition for electron impact on H_2* Chem.Phys.Lett. **43** 240

50.   Baltayan P. and Nedelec O. 1976 *Excitation of the $3p^3\Pi_u$ level of H_2 by electron impact; pressure effects* J.Quant.Spectrosc.Radiat.Tranfer **16** 207

51.   Bogdanova I.P., Efremova G.V., Lavrov B.P., Ostrovsky V.N., Ustimov V.I. and Yakovleva V.I. 1981 *Measurement of cross sections for electronic-rotational excitation of the $d^3\Pi_u^-,v,N$ states of ortohydrogen by electron impact* Opt.Spectrosc. **50** 63

52.   Lavrov B.P., Ostrovsky V.N. and Ustimov V.I. 1981 *Non-Franck-Condon transitions of molecules II. Semi-empirical approach: transitions in H_2.* J.Phys.B **14** 4701

53.   Lavrov B.P., Ostrovsky V.N. and Ustimov V.I. 1981 *Non-Franck-Condon transitions of molecules I. Theory.* J.Phys.B **14** 4389

54.   Chung S., Lin C.C. and Lee T.P. 1975 *Dissociation of the hydrogen molecule by electron impact* Phys.Rev.A **12** 1340

55.   Natalense A.P.P., Sartori C.S., Ferreira L.G. and Lima M.A.P. 1996 *Electronic excitation of H2 by electron impact using soft norm-conserving pseudopotentials.* Phys.Rev.A **54** 5435

56.   Lee Mu-Tao, Machado L.E., Brescansin L.M. and Meneses G.D. 1991 *Studies of vibronic excitations of H_2 by electron impact* J.Phys.B **24** 509

57.   Kiyoshima T., Sato S., Adamson S.O., Pazyuk E.A. and Stolyarov A.V. 1999 *Competition between predissociative and radiative decays in the $e^3\Sigma_u^+$ and $d^3\Pi_u^-$ states of H_2 and D_2* Phys.Rev.A **60** 4494

58.   Shevelko V.P. 1997 *Atoms and their Spectroscopic properties* (Spinger-Verlag Berlin Heidelberg) ISSN 0177-6495

59.   Wood R.W. 1920 *An extension of the Balmer series of hydrogen and spectroscopic phenomena of very long vacuum tubes* Roy.Soc.Proc. A **XCVII** 455

60.   Wood R.W. 1923 *Spontaneous Incandescence of Substances in Atomic Hydrogen Gas* Roy.Soc.Proc. A **CII** 1

61.   Ohl A. 1993 *Large Area Planar Microwaves Plasmas*, in: Microwave Discharges: Fundamentals and Applications, ed Ferreira C.M. and Moissan M. (New York: Plenum Press) p 205

62.   Lavrov B.P., Osiac M., Pipa A.V. and Röpcke J. 2003 *On the spectroscopic detection of neutral species in low-pressure plasma containing boron and hydrogen* Plasma Sources Sci. Technol. **12** 576

63.   Osiac M., Lavrov B.P. and Röpcke J. 2002 *Intensity distribution in R and P branches of (0-0) band of the $A^1\Pi \rightarrow X^1\Sigma^+$ electronic transition of the BH molecule and determination of the gas temperature in non-equilibrium plasmas* J.Quant.Spectrosc.Radiat.Transf. **74** 471

64.   Griem H.R. 1964 *Plasma Spectroscopy* McGraw-Hill Book Company

65.   Latyev L.N., Petrov V.A., Chehovskoi V.Ya. and Shestakov 1974 *Emission properties of solid material* ed. Sheindlin A.E. (Moscow: Energiya)

66.   Lang N. 2001 *Zu zeitabhängigen Anregungs- und Relaxationsphänomenen in wasserstoffhaltigen Mikrowellenplasmen* Inaugural dissertation, Uni-Greifswald.

67.   Lavrov B.P. and Otorbaev D.K. 1978 *Relation between rotational temperature and gas temperature in a low-pressure molecular plasma* Sov.Theh.Phys.Lett. **4** 574

68.   Lavrov B.P. and Tyutchev M.V. 1984 *Gas temperature measurements in non-equilibrium plasma from the intensities of H2 molecular bands* Acta Phys. Hung. **55** 411

69.   Drachev A.I. and Lavrov B.P. 1988 *Gas temperature determination from the intensity distribution in the rotational structure of diatomic-molecule bands excited by electron impact.* High Temp. **26** 129

70.   Jost W. 1952 *Diffusion in solids, liquids, gases* ed. Hutchinson E. (New York: Academic Press Inc.) p. 413-420

GLOSSARY

Definitions

Case 1   when the mixing of the population densities of the sublevels is extremely small and can be neglected. This situation is typical for gas-beam and crossed beam experiments to determine collision cross sections,

Case 2   when the mixing processes are dominant for the formation of the population density distribution among various fine structure sublevels. This situation could appear in plasmas with high enough densities of particles and fields.

Abbreviation

CARS	coherent anti-Stokes Raman scattering
CCC	convergent close coupling
CCD	charge coupling device
DC	direct current
EEDF	electron energy distribution function
LIF	laser induced fluorescence
MW	microwave
PC	personal computer
REMPI	resonance enhanced multi-photon ionization
RF	radio frequency
VUV	vacuum ultra violet
H_α, H_β, H_γ	Balmer $- \alpha$, β, γ ... spectral lines of atomic hydrogen
(2-2)Q1	spectral line of molecular hydrogen of the Fulcher-α band system, occurs in the transition $d^3\Pi_u^-$, v=2, N=1 $\rightarrow a^3\Sigma_g^+$ v=2, N=1

Symbols

A_{2lj}^{nlj}	spontaneous emission probability (Einstein coefficient) for a $nlj \rightarrow 2\tilde{l}\ \tilde{j}$ transition
A_{2l}^{nl}	spontaneous emission probability (Einstein coefficient) for a $nl \rightarrow 2\tilde{l}$ transition
$A_{n \rightarrow \tilde{n}}$	spontaneous emission probability (Einstein coefficient) for a $n \rightarrow \tilde{n}$ transition
$A_{a21}^{d^-21}$	probability of the $d^3\Pi_u^-$, v=2, N=1 $\rightarrow a^3\Sigma_g^+$ v=2, N=1 radiative transition of the hydrogen molecule
D	the degree of the dissociation of hydrogen
$d\lambda/dx$	linear dispersion of the spectrometer
$\Delta\Omega$	solid angle
E_N	energy of the rotational levels of the ground state of the hydrogen molecule
E_{X0N}	energy of the rotational levels of the $X\Sigma_g^+$, v=0, N level of the hydrogen molecule

ε	electron energy
ε_{ab}	the threshold energy for the $b \leftarrow a$ process
$\varepsilon(\lambda)$	plasma emissivity
$F(\varepsilon)$	EEDF normalized as $\int\limits_{0}^{\infty} F(\varepsilon)d\varepsilon = 1$.
$f(\varepsilon)$	excitation functions
$f_{d \leftarrow X}(\varepsilon)$	excitation function of the $d^3\Pi_u^-$ v = 2, N = 1 level of the hydrogen molecule
f	frequency
g_{nlj}	statistical weight of the nlj-sublevel of hydrogen atom
g_{nl}	statistical weight of the nj-sublevel of hydrogen atom
h	Planck constant
H_{NN}	Hönle-London factor
[H]	concentration of the hydrogen atoms
[H$_2$]	concentration of the hydrogen molecules
[H$_2$]$_{X01}$	population density of the $X^1\Sigma_g$ v = 0, N = 1 level of the hydrogen molecule
[H$_2$]$_{ortho}$	the concentrations of ortho-hydrogen
[H$_2$]$_{para}$	the concentrations of para-hydrogen
$\eta(T_g, T_0)$	ratio [H$_2$]$_{X01}$/[H$_2$] as function T_g and T_0
$\eta(T_g)$	ratio [H$_2$]$_{X01}$/[H$_2$] as function T_g assuming [H$_2$]$_{ortho}$/[H$_2$]$_{para}$ = 3
$I_{n \rightarrow 2}$	intensity of the Balmer line
$I_{3 \rightarrow 2}$, $I_{H\alpha}$	intensity of the H$_\alpha$ line
$I_{4 \rightarrow 2}$, $I_{H\beta}$	intensity of the H$_\beta$ line
I_{22Q1}	intensity of the (2-2)Q1 line of the Fulcher-α band system
$I_{2\tilde{i}j}^{nlj}$	intensity of a separate component of the fine structure of the Balmer line
I, i	electrical current
k	Boltzmann constant
$K_{BL}(\lambda)$	signal given by the detector from band lamp
$K_{PL}(\lambda)$	signal given by the detector from plasma
L_{PL}	length of the plasma column
l	linear dimension of a pixel
m	electron mass
n, l, j	principle, orbital and total angular momentum quantum numbers
$\tilde{n}, \tilde{l}, \tilde{j}$	principle, orbital and total angular momentum quantum numbers
N	rotational quantum number (total angular momentum, without considering the electron spin)
N_{nlj}	population density of the nlj - sublevel of the hydrogen atom
N_n	population density of the level n of the hydrogen atom
n_e	electron concentration
n_{X0N}	population density of a rovibronic sublevel of the ground state $X\Sigma_g^+$, $v=0$, N of molecular hydrogen
$n_{dv'N}$	population density of the $d^3\Pi_u^-$, v', N sublevel
p	pressure
P	power
R_{nlj}^+, R_{nlj}^-	total rates of population and depopulation of a nlj - sublevel of the hydrogen atom due to secondary processes.
$|R_e^{da}|^2$	square of the matrix element of the dipole moment

$q_{v'v'}$	Franck –Condon factor
$\rho(\lambda,T)$	emissivity coefficient of tungsten
$S(\lambda)$	sensitivity of the spectral system
$\sigma(\lambda)$	quantum efficiency of the detector
T	temperature of the tungsten band
T_e	electron temperature
T_e^{eff}	effective parameter of the approximation of the EEDF in the threshold region by a Maxwellian function.
T_g	gas temperature
T_0	temperature, which determine the ratio of concentrations of ortho- and para-hydrogen
T_{rot}^{X0}	rotational temperature of the electronic and vibrational ground state $X\Sigma_g^+$, v=0
τ_{nlj}	radiative lifetime of the nlj - sublevel of the hydrogen atom
τ_{nl}	radiative lifetime of the nl - sublevel of the hydrogen atom
τ_n	radiative lifetime of the level n of the hydrogen atom
τ_{d^-21}	radiative life time of the $d^3\Pi_u^-$, v=2, N=1 state
τ_{BL}	exposure time to the band lamp
τ_{PL}	exposure time to the plasma
v, v'	vibrational quantum numbers
$v_{av'N}^{dv'N}$	wavenumber of the spectral line of the radiative transition $d^3\Pi_u^-$, v', N=1 $\rightarrow$ $a^3\Sigma_g^+$ v', N

Rate coefficients

$K^{ex}(F(\varepsilon)\vert b \leftarrow a)$	the excitation rate coefficients for the transition $b \leftarrow a$ function of EEDF
$K^{ex}(T_e \vert b \leftarrow a)$	the excitation rate coefficients for the transition $b \leftarrow a$ function of T_e
$K_{dir}^{ex}(F(\varepsilon)\vert nlj \leftarrow 1s)$	the excitation rate coefficients for the direct electron impact excitation of nlj – sublevel of the hydrogen atom as function of EEDF
$K_{dir}^{ex}(F(\varepsilon)\vert nl \leftarrow 1s)$	the excitation rate coefficients for the direct electron impact excitation of nl – sublevel of the hydrogen atom as function of EEDF
$K_{dir}^{ex}(F(\varepsilon)\vert n \leftarrow 1s)$	the excitation rate coefficients for the direct electron impact excitation of level n of the hydrogen atom as function of EEDF
$K_{dis}^{ex}(F(\varepsilon)\vert nlj \leftarrow X)$	the excitation rate coefficients for the dissociative electron impact excitation of nlj – sublevel of the hydrogen atom as function of EEDF
$K_{dis}^{ex}(F(\varepsilon)\vert nl \leftarrow X)$	the excitation rate coefficients for the dissociative electron impact excitation of nl – sublevel of the hydrogen atom as function of EEDF
$K_{dis}^{ex}(F(\varepsilon)\vert n \leftarrow X)$	the excitation rate coefficients for the dissociative electron impact excitation of level n of the hydrogen atom as function of EEDF

$K^{em}\left(F(\varepsilon)|n\to 2\right)$ — emission rate coefficients of Balmer lines

$K^{em}_{dir}\left(F(\varepsilon)|n\to 2\right)$ — emission rate coefficient for direct excitation by electron impact of Balmer lines

$K^{em}_{dis}\left(F(\varepsilon)|n\to 2\right)$ — emission rate coefficients dissociative excitation by electron impact of Balmer lines

K^{em1}, K^{em2} — emission rate coefficients corresponding to *case 1* and *case 2*

$K^{em2}_{dir}\left(H_\alpha\right)$ — emission rate coefficients of direct excitation of H_α line by electron impact, corresponding to case 2

$K^{em}_{dir}\left(T^{eff}_e \mid H_\alpha\right)$ — emission rate coefficients of direct excitation of H_α line by electron impact as a function of T_e^{eff}

$K^{em}_{dis}\left(T^{eff}_e \mid H_\alpha\right)$ — emission rate coefficients of dissociative excitation of H_α line by electron impact as a function of T_e^{eff}

$K^{em}_{dir}\left(T^{eff}_e \mid H_\beta\right)$ — emission rate coefficients of direct excitation of H_β line by electron impact as a function of T_e^{eff}

$K^{em}_{dis}\left(T^{eff}_e \mid H_\beta\right)$ — emission rate coefficients of dissociative excitation of H_α line by electron impact as a function of T_e^{eff}

$K^{ex}_{mol}\left(F(\varepsilon)| d^-21 \leftarrow X01\right)$ — excitation rate coefficient for the electron impact of the $d^3\Pi_u^-$, v=2, N=1 state of molecular hydrogen.

$K^{d v'N}_{X0N}$ — excitation rate coefficient for the electron impact of molecular hydrogen, transition $X^1\Sigma_g$ v, N $\to$ $d^3\Pi_u^-$, v', N

Cross sections

$\sigma^{ex}_{b\leftarrow a}(\varepsilon)$ — the cross section of the electron-impact excitation of the b level

$\sigma^{ex}_{nl\leftarrow 1s}(\varepsilon)$ — the excitation cross section of nl-sublevel of hydrogen atom for direct electron impact excitation from the ground state of the atom

$\sigma^{ex}_{nl\leftarrow X}(\varepsilon)$ — the excitation cross section of nl-sublevel of hydrogen atom for dissociative electron impact excitation from the ground state of the molecule

$\sigma_{1s\to 3s}$, $\sigma_{1s\to 3p}$, $\sigma_{1s\to 3d}$ — cross sections of hydrogen atom 3s-, 3p-, and 3d- sublevels excitation by electron impact

$\sigma^{em}(\varepsilon)$ — emission cross section

$\sigma^{em2}(\varepsilon|n\to 2)$ — emission cross section of Balmer lines

$\sigma^{em1}_{dir}(\varepsilon \mid n\to 2)$ — emission cross section of Balmer lines for direct electron impact excitation corresponding to *case 1*

$\sigma^{em2}_{dir}(\varepsilon \mid n\to 2)$ — emission cross section of Balmer lines for direct electron impact excitation corresponding to *case 2*

$\sigma^{em1}_{dis}(\varepsilon \mid n\to 2)$ — emission cross section of Balmer lines for dissociative electron impact excitation corresponding to *case 1*

$\sigma_{dis}^{em2}(\varepsilon \mid n \to 2)$ emission cross section of Balmer lines for dissociative electron impact excitation corresponding to *case 2*

$\sigma_{dir}^{em}(H_\alpha)$, $\sigma_{dir}^{em}(H_\beta)$ emission cross sections of direct excitation of the H_α and H_β lines by electron impact

$\sigma_{dir}^{em1}(H_\alpha)$, $\sigma_{dir}^{em1}(H_\beta)$ emission cross sections of direct excitation of the H_α and H_β lines by electron impact corresponding to *case 1*

$\sigma_{d^-21\leftarrow X01}^{max}$ value of the excitation cross section of the $d^3\Pi_u^-,(v = 2, N = 1)$ level by electron impact from the ground state $X^1\Sigma_g^+$ $v = 0, N = 1$ of hydrogen molecule in maximum

$\sigma_{dis}^{em1}(50eV \mid 3 \to 2)$ value of emission cross section of H_α line for dissociative electron impact excitation corresponding to *case 1* at electron energy 50 eV

APPENDIX

Table 1. Transition probabilities of spontaneous emission and radiative life times of fine structure sublevels of the hydrogen atom from ref. [18].

Term nl	g_{nl}	τ_{nl} 10^{-9} sec	$A_{\tilde{n}\tilde{l}}^{nl}$, 10^{7} sec^{-1}							
			$2^2P_{1/2}$	$2^2P_{3/2}$						
$3^2S_{1/2}$	2	160	0.21	0.42						
			$1^2S_{1/2}$	$2^2S_{1/2}$						
$3^2P_{1/2}$	2	5.4	16.4	2.23						
			$1^2S_{1/2}$	$2^2S_{1/2}$						
$3^2P_{3/2}$	4	5.4	16.4	2.23						
			$2^2P_{1/2}$	$2^2P_{3/2}$						
$3^2D_{3/2}$	4	15.6	5.36	1.07						
			$2^2P_{3/2}$							
$3^2D_{5/2}$	6	15.6	6.43							
			2P	3P						
4S	2	230	0.25	0.18						
			1S	2S	3S	3D				
4P	6	12.4	6.8	0.95	0.30	0.03				
			2P	3P						
4D	10	36.5	2.04	0.70						
			3D							
4F	14	73	1.37							
			2P	3P	4P					
5S	2	360	0.12	0.08	0.06					
			1S	2S	3S	3D	4S	4D		
5P	6	24	3.4	0.49	0.16	0.01	0.07	0.02		
			2P	3P	4P	4F				
5D	10	70	0.94	0.34	0.14	0.00				
			3D	4D						
5F	14	140	0.45	0.26						
			4F							
5G	18	235	0.42							
			2P	3P	4P	5P				
6S	2	570	0.07	0.051	0.035	0.017				
			1S	2S	3S	3D	4S	4D	5S	5D
6P	6	41	1.95	0.029	0.096	0.007	0.045	0.009	0.021	0.010
			2P	3P	4P	4F	5P	5F		
6D	10	126	0.48	0.187	0.086	0.002	0.040	0.004		
			3D	4D	5D	5G				
6F	14	243	0.210	0.0129	0.072	0.001				
			4F	5F						
6G	18	405	0.137	0.110						
			5G							
6H	22	610	0.164							

Table 2. Comparison of "equilibrium" transition probabilities of spontaneous emission and radiative life times for levels with n=2-4, from data of refs. [18] and [22]

Vainshtein, et.al. [22]						E.U.Condon, G.H. Shortley [18]					
τ_n 10^{-7}s	$\tau_n A_{n\to2}$	$A_{n\to\tilde{n}}$, 10^{7}s				τ_n 10^{-7}s	$\tau_n A_{n\to2}$	$A_{n\to\tilde{n}}$, 10^{7}s			
		$\tilde{n}$	1	2	3			$\tilde{n}$	1	2	3
		n						n			
0.0213	1	2	47.0			0.0213	1	2	46.9		
0.100	0.442	3	5.57	4.41		0.101	0.445	3	5.47	4.39	
0.331	0.279	4	1.28	0.842	0.899	0.336	0.279	4	1.275	0.831	0.869

Table 3. Emission cross sections of H_α and H_β lines ($10^{-18} cm^2$).

T_e, eV	Direct excitation				Dissociative excitation			
	H_α		H_β		H_α		H_β	
	Case 1	Case 2	Case 1	Case 2	Case 1	Case 2	Case 1	Case 2
12.09	0	0	--	--	--	--	--	--
12.75	2.18	1.72	0	0	--	--	--	--
14	6.32	4.97	1.33	0.930	--	--	--	--
14.5	6.22	4.99	1.86	1.30	--	--	--	--
15	6.12	5.01	1.72	1.28	--	--	--	--
16	5.93	5.07	1.70	1.27	--	--	--	--
16.57	5.83	5.10	1.69	1.27	0	0	--	--
17.23	5.73	5.14	1.68	1.27	0.110	0.0515	0	0
18	5.61	5.18	1.67	1.27	0.243	0.112	0.0277	0.00246
19	5.47	5.24	1.66	1.27	0.412	0.190	0.0637	0.00564
20	5.34	5.29	1.64	1.27	0.584	0.269	0.0648	0.00569
21	5.23	5.35	1.63	1.27	0.590	0.271	0.0658	0.00576
22	5.13	5.40	1.62	1.28	0.594	0.273	0.0667	0.00583
23	5.05	5.46	1.60	1.29	0.598	0.274	0.0677	0.00589
24	4.99	5.52	1.59	1.30	0.601	0.276	0.0687	0.00596
25	4.92	5.57	1.58	1.32	0.605	0.278	0.0697	0.00603
26	4.87	5.63	1.56	1.34	0.607	0.279	0.0707	0.00609
27	4.83	5.68	1.55	1.34	0.627	0.288	0.0717	0.00616
28	4.79	5.74	1.54	1.35	0.648	0.298	0.0790	0.00650
29	4.75	5.79	1.52	1.35	0.667	0.307	0.0929	0.00755
30	4.71	5.86	1.51	1.36	0.686	0.315	0.100	0.00828
31	4.67	5.86	1.50	1.36	0.704	0.324	0.107	0.00897
32	4.63	5.85	1.48	1.37	0.722	0.332	0.114	0.00963
33	4.59	5.85	1.47	1.37	0.738	0.339	0.121	0.0103
34	4.55	5.84	1.45	1.38	0.754	0.347	0.127	0.0108
35	4.51	5.84	1.44	1.38	0.769	0.354	0.133	0.0114
36	4.47	5.83	1.43	1.38	0.783	0.360	0.138	0.0119
37	4.43	5.82	1.41	1.38	0.797	0.366	0.144	0.0124
38	4.39	5.82	1.40	1.38	0.810	0.372	0.149	0.0129
39	4.36	5.81	1.39	1.38	0.822	0.378	0.154	0.0133
40	4.32	5.81	1.37	1.38	0.833	0.383	0.158	0.0136
42	4.24	5.79	1.35	1.38	0.849	0.390	0.167	0.0143
44	4.16	5.78	1.32	1.37	0.865	0.397	0.174	0.0148
50	3.92	5.74	1.24	1.34	0.908	0.417	0.187	0.0156
55	3.74	5.72	1.20	1.33	0.939	0.432	0.193	0.0161
60	3.61	5.60	1.15	1.30	0.966	0.444	0.199	0.0165
70	3.37	5.37	1.07	1.24	1.01	0.463	0.206	0.0169
80	3.14	5.14	0.997	1.19	1.03	0.474	0.206	0.0169
90	2.93	4.93	0.927	1.13	1.04	0.477	0.199	0.0164
100	2.74	4.73	0.863	1.09	1.00	0.461	0.191	0.0156
120	2.40	4.35	0.750	0.997	0.940	0.432	0.176	--
140	2.12	4.02	0.659	0.917	0.880	0.405	0.162	--
160	1.90	3.73	0.591	0.849	0.824	0.379	0.149	--
180	1.74	3.47	0.545	0.791	0.770	0.355	0.137	--
200	1.65	3.27	0.519	0.744	0.720	0.331	0.126	--
250	1.42	2.91	0.439	0.642	0.608	--	0.102	--
300	1.23	2.59	0.374	0.557	0.514	--	0.0838	--
350	1.07	2.31	0.322	0.490	0.439	--	0.0717	--
400	0.939	2.07	0.283	0.439	0.384	--	0.0655	--
450	0.840	1.87	0.257	0.407	0.3475	--	0.0592	--
500	0.775	1.72	0.244	0.391	0.325	--	0.0535	--
600	0.708	1.58	0.225	0.358	0.271	--	0.04423	--
800	0.569	1.29	0.181	0.292	0.212	--	0.0338	--
1000	0.430	1.00	0.137	0.227	0.178	--	0.0271	--

Table 4. Rate coefficients (cm^3s^{-1}) of direct and dissociative excitation of the H_α and H_β lines by electron impact, calculated using a Maxwell electron energy distribution function at various electron temperatures T_e. The calculated data are given for the case when redistribution of populations can be neglected (*case 1*) and in the case of a Boltzmann distribution over the fine structure sublevels (*case 2*).

T_e, eV	Direct excitation				Dissociative excitation			
	H_α		H_β		H_α		H_β	
	Case 1	*Case 2*	*Case 1*	*Case 2*	*Case 1*	*Case 2*	*Case 1*	*Case 2*
0.5	6.29E-20	4.97E-20	5.60E-21	3.93E-21	5.71E-25	2.63E-25	3.26E-26	2.90E-27
1	1.44E-14	1.15E-14	2.39E-15	1.70E-15	1.29E-17	5.94E-18	1.25E-18	1.10E-19
1.5	8.62E-13	7.03E-13	1.75E-13	1.26E-13	3.84E-15	1.77E-15	4.14E-16	3.65E-17
2	6.53E-12	5.43E-12	1.46E-12	1.07E-12	6.65E-14	3.06E-14	7.44E-15	6.55E-16
2.5	2.17E-11	1.84E-11	5.16E-12	3.82E-12	3.68E-13	1.69E-13	4.20E-14	3.68E-15
3	4.77E-11	4.14E-11	1.19E-11	8.86E-12	1.15E-12	5.28E-13	1.34E-13	1.17E-14
3.5	8.33E-11	7.38E-11	2.14E-11	1.61E-11	2.60E-12	1.19E-12	3.08E-13	2.68E-14
4	1.26E-10	1.14E-10	3.32E-11	2.53E-11	4.79E-12	2.20E-12	5.79E-13	5.03E-14
4.5	1.73E-10	1.60E-10	4.66E-11	3.58E-11	7.72E-12	3.55E-12	9.54E-13	8.26E-14
5	2.23E-10	2.09E-10	6.10E-11	4.73E-11	1.13E-11	5.21E-12	1.43E-12	1.24E-13
5.5	2.74E-10	2.61E-10	7.59E-11	5.94E-11	1.56E-11	7.15E-12	2.01E-12	1.73E-13
6	3.24E-10	3.14E-10	9.10E-11	7.19E-11	2.03E-11	9.32E-12	2.68E-12	2.31E-13
6.5	3.74E-10	3.68E-10	1.06E-10	8.45E-11	2.54E-11	1.17E-11	3.43E-12	2.95E-13
7	4.22E-10	4.21E-10	1.21E-10	9.71E-11	3.09E-11	1.42E-11	4.26E-12	3.65E-13
7.5	4.68E-10	4.74E-10	1.35E-10	1.10E-10	3.67E-11	1.69E-11	5.15E-12	4.42E-13
8	5.12E-10	5.26E-10	1.49E-10	1.22E-10	4.26E-11	1.96E-11	6.11E-12	5.23E-13
8.5	5.55E-10	5.77E-10	1.62E-10	1.34E-10	4.87E-11	2.24E-11	7.11E-12	6.08E-13
9	5.95E-10	6.27E-10	1.75E-10	1.46E-10	5.50E-11	2.53E-11	8.16E-12	6.96E-13
9.5	6.34E-10	6.75E-10	1.87E-10	1.57E-10	6.13E-11	2.82E-11	9.25E-12	7.88E-13
10	6.70E-10	7.22E-10	1.98E-10	1.68E-10	6.77E-11	3.11E-11	1.04E-11	8.81E-13
11	7.38E-10	8.12E-10	2.20E-10	1.89E-10	8.06E-11	3.71E-11	1.27E-11	1.08E-12
12	7.98E-10	8.96E-10	2.40E-10	2.09E-10	9.34E-11	4.30E-11	1.50E-11	1.28E-12
13	8.53E-10	9.76E-10	2.58E-10	2.28E-10	1.06E-10	4.88E-11	1.74E-11	1.48E-12
14	9.03E-10	1.05E-09	2.74E-10	2.45E-10	1.19E-10	5.46E-11	1.98E-11	1.69E-12
16	9.88E-10	1.18E-09	3.02E-10	2.77E-10	1.43E-10	6.57E-11	2.46E-11	2.09E-12
18	1.06E-09	1.30E-09	3.25E-10	3.05E-10	1.66E-10	7.64E-11	2.92E-11	2.48E-12
20	1.12E-09	1.41E-09	3.45E-10	3.29E-10	1.88E-10	8.64E-11	3.36E-11	2.85E-12
25	1.23E-09	1.63E-09	3.81E-10	3.79E-10	2.37E-10	1.09E-10	4.34E-11	3.69E-12
30	1.30E-09	1.79E-09	4.05E-10	4.18E-10	2.78E-10	1.28E-10	5.17E-11	4.39E-12
40	1.38E-09	2.03E-09	4.33E-10	4.71E-10	3.43E-10	1.58E-10	6.42E-11	5.46E-12
50	1.42E-09	2.18E-09	4.46E-10	5.05E-10	3.89E-10	1.79E-10	7.28E-11	6.19E-12
75	1.45E-09	2.39E-09	4.53E-10	5.48E-10	4.55E-10	2.08E-10	8.40E-11	7.14E-12
100	1.44E-09	2.48E-09	4.52E-10	5.64E-10	4.83E-10	2.22E-10	8.80E-11	7.48E-12

Table 5. Excitation function of the $d\,^3\Pi_u^-$, (v=2, N=1) level of the hydrogen molecule by electron impact, normalized on its maximum.

$\varepsilon,$ eV	$f_{d\leftarrow X}(\varepsilon)$	$\varepsilon,$ eV	$f_{d\leftarrow X}(\varepsilon)$	$\varepsilon,$ eV	$f_{d\leftarrow X}(\varepsilon)$	$\varepsilon,$ eV	$f_{d\leftarrow X}(\varepsilon)$
14.4	0	27	0.398	42	0.119	100	0.00989
15	0.500	28	0.363	44	0.101	110	0.00735
15.6	1.00	29	0.330	46	0.0878	120	0.00578
16	0.974	30	0.299	48	0.0779	140	0.00363
17	0.907	31	0.280	50	0.0713	160	0.00241
18	0.844	32	0.260	55	0.0585	180	0.00174
19	0.784	33	0.242	60	0.0473	200	0.00122
20	0.726	34	0.225	65	0.0378	250	6.41E-4
21	0.671	35	0.208	70	0.0298	300	3.85E-4
22	0.618	36	0.193	75	0.0235	350	2.26E-4
23	0.569	37	0.178	80	0.0188	400	1.57E-4
24	0.522	38	0.164	85	0.0157	500	7.95E-5
25	0.478	39	0.152	90	0.0136	600	4.66E-5
26	0.437	40	0.140	95	0.0115	700	3.01E-5

Table 6. Rate coefficients K^{ex}_{mol} ($cm^3 s^{-1}$) for the excitation of the $d^3\Pi_u^-$, v=2, N=1 level of molecular hydrogen by electron impact from the ground electronic-vibronic state, calculated using a Maxwellian electron energy distribution function at various electron temperatures T_e multiplied with the radiative life time of this state $\tau_{d\,21}^-$ and the radiative transition probabilities $A^{d^-21}_{a21}$ of the transition $d^3\Pi_u^-$, v=2, N=1 $\rightarrow$ $a^3\Sigma_g^+$, v=2, N=1

T_e, eV	$A^{d^-21}_{a21}\tau_{d^-21}K^{ex}_{mol}$	T_e, eV	$A^{d^-21}_{a21}\tau_{d^-21}K^{ex}_{mol}$	T_e, eV	$A^{d^-21}_{a21}\tau_{d^-21}K^{ex}_{mol}$
0.5	1.06E-22	6	1.94E-11	14	4.19E-11
1	2.10E-16	6.5	2.22E-11	16	4.26E-11
1.5	2.45E-14	7	2.48E-11	18	4.24E-11
2	2.54E-13	7.5	2.73E-11	20	4.18E-11
2.5	1.00E-12	8	2.94E-11	25	3.91E-11
3	2.46E-12	8.5	3.14E-11	30	3.58E-11
3.5	4.58E-12	9	3.32E-11	40	2.99E-11
4	7.21E-12	9.5	3.47E-11	50	2.51E-11
4.5	1.02E-11	10	3.61E-11	75	1.72E-11
5	1.33E-11	11	3.83E-11	100	1.27E-11
5.5	1.64E-11	12	4.00E-11		

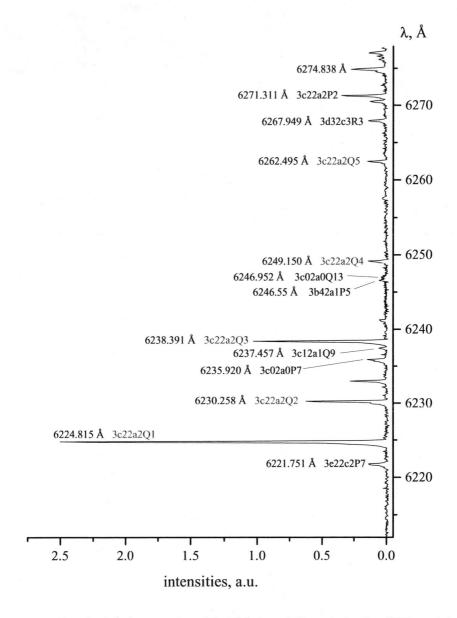

Example of a hydrogen spectrum of the RF discharge, in H_2, $p = 1$ mbar, $T_g = 400$ K recorded with a 0.5 m spectrometer and a grating of 2400 g/mm

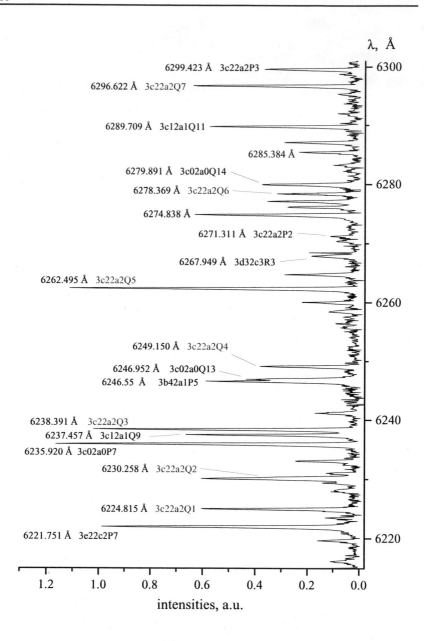

Example of a hydrogen spectrum of the MW discharge, in H_2, p = 50 mbar, T_g = 2000K, recorded with a 0.5 m spectrometer and a grating of 2400 g/mm

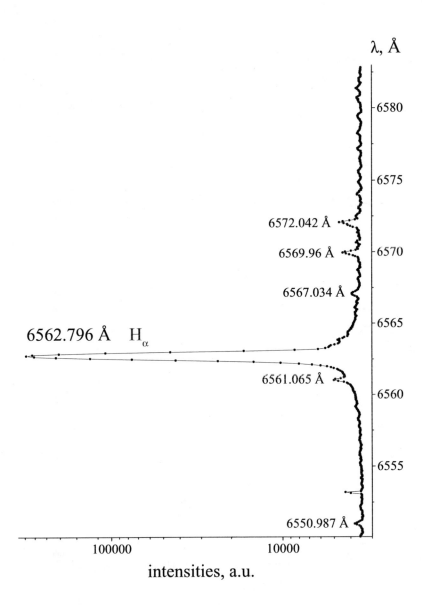

Example of a hydrogen spectrum of the MW discharge, in H_2 + Ar + B_2H_6, p = 2.5 mbar, T_g = 1000K, recorded with a 0.5 m spectrometer and a grating of 2400 g/mm

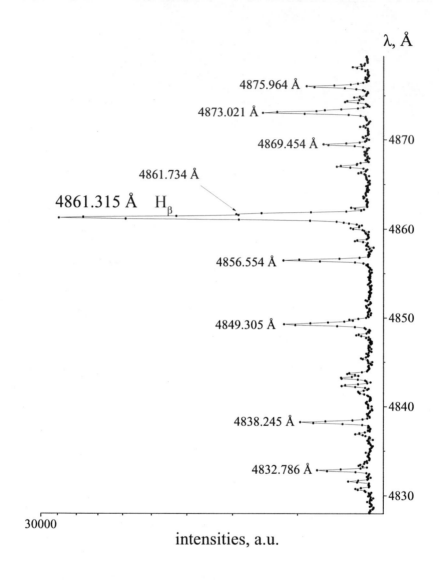

Example of a hydrogen spectrum of the MW discharge, in H_2 + Ar + B_2H_6, p = 2.5 mbar, gas temperature 1000K, recorded with a 0.5 m spectrometer and a grating of 2400 g/mm

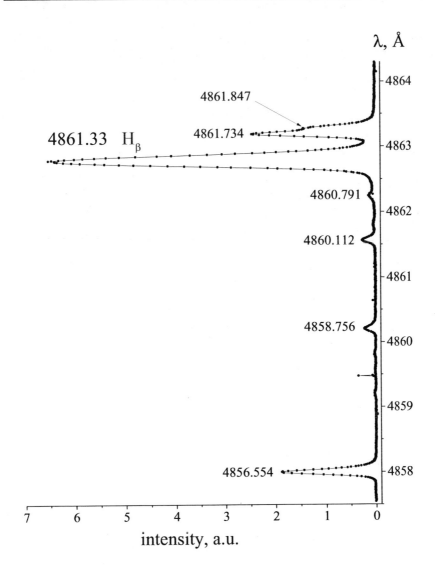

Example of a hydrogen spectrum of the DC-arc discharge, lamp DVS-25, in H_2, p = 8 mbar, T_g = 1000K, recorded with a 1 m – double spectrometer and a grating of 1800 g/mm

Curriculum vitae

Andrei V. Pipa

Date of birth: January 31, 1977
Place of birth: Murmansk
Citizenship: Russia
Marital Status: not married

Mailing addresses

Office: Institut für Niedertemperatur Plasmaphysik Greifswald
        Friedrich-Ludwig-Jahn-Str. 19
        17489 Greifswald
        Germany
E-mail: pipa@pisem.net; pipa@inp-greifswald.de
Phone: +49 (0) 3834 554 431 FAX: +49 (0) 3834 554 301

Education and degrees

Bachelor of Physics (Optics and Spectroscopy) St.-Petersburg State University, Russia, 1999.
Master of Physics (Optics and Spectroscopy) St.-Petersburg State University, Russia, 2001.

Employment

Sep. 1994 - Jun. 1999	Student Department of Optics, Faculty of Physics, St.-Petersburg State University, Russia
Jul. 1999 - Jun. 2001	Master of Science Student Department of Optics, Faculty of Physics, St.-Petersburg State University, Russia
Sep. 2000 - Feb. 2001	DAAD/IAEESTE practice student Institute of Low-Temperature Plasma Physics Greifswald, Germany
March 2001 – Sep. 2004	scientific assistant in Institute of Low Temperature Plasma Physics Greifswald, Germany

Scientific interests

Spectroscopy of low temperature plasma; determination of dissociation degree, gas temperature; studies of cross sections and rate coefficients. The object of the investigations in most cases contains hydrogen atoms and molecules. Additional interest in Tunable Diode Laser Absorption Spectroscopy (TDLAS) applications.

List of Publications

1. B. P. Lavrov and A. V. Pipa
Account of the Fine Structure of Hydrogen Atom Levels in the Emission Cross Sections of the Balmer Lines Excited by Electron Impact in Gases and Plasma
Opt. Spectrosc. **92** (2002) 647-657.

2. F. Hempel, J. Röpcke, A. Pipa and P. B. Davies
Infrared laser spectroscopy of the CN free radical in a methane-nitrogen-hydrogen plasma
Molecular Physics, **101** (2003) 589-594

3. B. P. Lavrov, M. Osiac, A. V. Pipa and J. Röpcke
On the spectroscopic detection of neutral species in a low-pressure plasma containing boron and hydrogen
Plasma Sources Sci. Technol. **12** (2003) 576-589

Contributions

1. A. Pipa, N. Lang, B. Lavrov, and J. Röpcke.
On Spectroscopic Determination of Hydrogen Dissociation Degree in Pulse Modulated Microwave Discharge.
ECAMP VII – Frühjahrstagung Plasmaphysik der DPG (Berlin 2001) Verhandlg. d. DPG, 5/2001, p. 169

3. B. P. Lavrov and A. V. Pipa and J. Roepcke
Electron Impact Excitation of Atoms and Spectroscopic Determination of the Dissociation Degree of Hydrogen in Non-Equilibrium Plasmas
Second Conference on the Elementary Processes in Atomic Systems (Gdansk 2002) Book of Abstracts, p. 79

4. A. V. Pipa, B. P. Lavrov and J. Röpcke
On spectroscopic determination of dissociation degree of hydrogen in Arc discharge
Frühjahrstagung Plasmaphysik der DPG (Aachen 2003) Verhandlungen 4/2003, p. 52

5. B. P. Lavrov, A. V. Pipa and J. Röpcke
On spectroscopic determination of hydrogen dissociation degree in non-equilibrium plasmas of DC, RF and microwave discharges
Frontiers in Low Temperature Plasma Diagnostics V (Villaggio Cardigliano 2003) Proc. p. 196-9

6. B. P. Lavrov, M. Osiac, A. V. Pipa and J. Röpcke
On the spectroscopic detection of neutral species in a low-pressure plasma containing boron and hydrogen
Frontiers in Low Temperature Plasma Diagnostics V (Villaggio Cardigliano 2003) Proc. p.152-5,

7. A. V. Pipa, B. P. Lavrov and J. Röpcke
On spectroscopic determination of dissociation degree of hydrogen in a RF discharge
ICPIG XXVI, (Greifswald 2003) Proc.1, p. 161,

8. J. Röpcke, P. Davies, F. Hempel, A. Pipa and G. Stancu
On recent progress in monitoring transient molecular species in microwave plasmas
V[th] International Workshop on Microwave Discharges: Fundamentals and Applications
(Greifswald 2003), Abstracts and Program, p. 95

9. A.V. Pipa, G. Lombardi, X. Duten, K. Hassouni and J. Röpcke
On the influence of argon addition on microwave hydrogen plasmas obtained under moderate pressure conditions.
Frühjahrstagung Plasmaphysik der DPG (Kiel 2004) Verhandlg. d. DPG, 4/2004, p. 39